THE GOD EQUATION

DR. MICHIO KAKU

PROFESSOR OF THEORETICAL PHYSICS

CITY UNIVERSITY OF NEW YORK

THE GOD EQUATION

The Quest for a Theory of Everything

DOUBLEDAY  NEW YORK

DOUBLEDAY and the portrayal of an anchor with a
dolphin are registered trademarks of Penguin Random
House LLC.

Jacket image by Eugen Domentean / Shutterstock
Jacket design by Michael J. Windsor
Book design by Michael Collica
Illustrations by Mapping Specialists Ltd.

Library of Congress Cataloging-in-Publication Data
Names: Kaku, Michio, author.
Title: The God equation : the quest for a theory of
everything / Michio Kaku.
Description: New York : Doubleday, [2021]
Identifiers: LCCN 2020034243 | ISBN 9780385542746
(hardcover) | ISBN 9780385542753 (ebook)
Subjects: LCSH: Cosmology. | Big bang theory.
Classification: LCC QB981 .K133 2021 | DDC 523.1—dc23
LC record available at https://lccn.loc.gov/2020034243

MANUFACTURED IN THE UNITED STATES OF AMERICA

5 7 9 10 8 6

First Edition

To my loving wife, Shizue, and my daughters,
Dr. Michelle Kaku and Alyson Kaku

CONTENTS

THE GOD EQUATION

INTRODUCTION TO THE FINAL THEORY

t was to be the final theory, a single framework that would unite all the forces of the cosmos and choreograph everything from the motion of the expanding universe to the most minute dance of subatomic particles. The challenge was to write an equation whose mathematical elegance would encompass the whole of physics.

Some of the most eminent physicists in the world embarked upon this quest. Stephen Hawking even gave a talk with the auspicious title "Is the End in Sight for Theoretical Physics?"

If such a theory is successful, it would be science's crowning achievement. It would be the holy grail of physics, a single formula from which, in principle, one could derive all other equations, starting from the Big Bang and moving to the end of the universe. It would be the end product of two thousand years of scientific investigation ever since the ancients asked the question, "What is the world made of?

It is a breathtaking vision.

EINSTEIN'S DREAM

I first came across the challenge this dream posed as a child of eight. One day, the newspapers announced that a great scientist had just died. There was an unforgettable picture in the paper.

It was an image of his desk, with an open notebook. The caption announced that the greatest scientist of our time could not finish the work he had started. I was fascinated. What could possibly be so hard that even the great Einstein could not solve it?

That book contained his unfinished theory of everything, what Einstein called the unified field theory. He wanted an equation, perhaps no more than one inch long, that would allow him to, in his words, "read the mind of God."

Not fully appreciating the enormity of this problem, I decided to follow in the footsteps of this great man, and hoped to play a small role in finishing his quest.

But many others have also tried and failed. As Princeton physicist Freeman Dyson once said, the road to the unified field theory is littered with the corpses of failed attempts.

Today, however, many leading physicists believe that we are finally converging on the solution.

The leading (and to my mind, only) candidate is called string theory, which posits the universe was not made of point particles but of tiny vibrating strings, with each note corresponding to a subatomic particle.

If we had a microscope powerful enough, we could see that electrons, quarks, neutrinos, etc. are nothing but vibrations on minuscule loops resembling rubber bands. If we pluck the rubber band enough times and in different ways, we eventually create all the known subatomic particles in the universe. This means that all the laws of physics can be reduced to the harmonies of these strings. Chemistry is the melodies one can play on them. The universe is a symphony. And the mind of God, which Einstein eloquently wrote about, is cosmic music resonating throughout space-time.

This is not just an academic question. Each time scientists have unraveled a new force, it has changed the course of civilization and altered the destiny of humanity. For example, Newton's discovery of the laws of motion and gravity laid the groundwork for the machine age and the Industrial Revolution. Michael Faraday and James Clerk Maxwell's explanation of electricity and magnetism paved the way for the illumination of our cities and gave us powerful electric motors and generators as well as instantaneous communication via TV and radio. Einstein's $E = mc^2$ explained the power of the stars and helped to unravel the nuclear force. When Erwin Schrödinger, Werner Heisenberg, and others unlocked the secrets of the quantum theory, they gave us the high-tech revolution of today, with supercomputers, lasers, the internet, and all the fabulous gadgets in our living rooms.

Ultimately, all the wonders of modern technology owe their origin to the scientists who gradually discovered the fundamental forces of the world. Now, scientists may be converging on the theory that unifies these four forces of nature—gravity, the electromagnetic force, and the strong and weak nuclear forces—into a single theory. Ultimately, it may answer some of the deepest mysteries and questions in all of science, such as:

- What happened before the Big Bang? Why did it bang in the first place?
- What lies on the other side of a black hole?
- Is time travel possible?
- Are there wormholes to other universes?
- Are there higher dimensions?
- Is there a multiverse of parallel universes?

This book is about the quest to find this ultimate theory and all the bizarre twists and turns of what is undoubtedly one of the strangest chapters in the history of physics. We will review all the previous revolutions, which have given us our technological marvels, starting with the Newtonian revolution, leading up to the mastery of the electromagnetic force, the development of relativity and the quantum theory, and the string theory of today. And we will explain how this theory may also unravel the deepest mysteries of space and time.

AN ARMY OF CRITICS

However, hurdles remain. For all the excitement generated by string theory, the critics have been keen to point out its defects. And after all the hype and frenzy, real progress has stalled.

The most glaring problem is that, for all the flattering press extolling the beauty and complexity of the theory, we have no solid, testable evidence. Once, it was hoped that the Large Hadron Collider (LHC) outside Geneva, Switzerland, the biggest particle accelerator in history, would find concrete evidence for the final theory, but this has remained elusive. The LHC was able to find the Higgs boson (or the God particle), but this particle was only a tiny missing piece of the final theory.

Although ambitious proposals have been made for an even more powerful successor to the LHC, there is no guarantee that these costly machines will find anything at all. No one knows for certain at what energy we will find new subatomic particles that could verify the theory.

But perhaps the most important criticism of string theory is that it predicts a multiverse of universes. Einstein once said that the key question was: Did God have a choice in making the universe? Is the universe unique? String theory by itself is unique, but it probably has an infinite number of solutions. Physicists call this the landscape problem—the fact that our universe may be just one solution among an ocean of other equally valid ones. If our universe is one of many possibilities, then which one is ours? Why do we live in this particular universe and not another? What, then, is the predictive power of string theory? Is it a theory of everything or a theory of anything?

I admit I have a stake in this search. I have been working on string theory since 1968, ever since it emerged accidentally, unannounced, and totally unexpected. I have seen the remarkable evolution of the theory that developed from a single formula into a discipline with a whole library's worth of research papers. Today, string theory forms the basis of much of the research being done in the world's leading laboratories. This book will hopefully give you a balanced, objective analysis of string theory's breakthroughs and limitations.

It will also explain why this quest has seized the imagination of the world's top scientists, and why this theory has generated so much passion and controversy.

1

UNIFICATION—THE ANCIENT DREAM

Gazing at the magnificent splendor of the night sky, surrounded by all the brilliant stars in the heavens, it is easy to be overwhelmed by its sheer, breathtaking majesty. Our concerns turn to some of the most mysterious questions of all.

Is there a grand design to the universe?

How do we make sense of a seemingly senseless cosmos?

Is there a rhyme and reason to our existence, or is it all pointless?

I am reminded of the poem by Stephen Crane:

A man said to the universe:

"Sir, I exist!"

"However," replied the universe,

"The fact has not created in me a sense of obligation."

The Greeks were among the first to make serious

attempts to sort through the chaos of the world around us. Philosophers like Aristotle believed that everything could be reduced to a mixture of four fundamental ingredients: earth, air, fire, and water. But how do these four elements give rise to the rich complexity of the world?

The Greeks proposed at least two answers to this question. The first was given by the philosopher Democritus, even before Aristotle. He believed that everything could be reduced to tiny, invisible, indestructible particles he called atoms (meaning "indivisible" in Greek). The critics, however, pointed out that direct evidence for atoms was impossible to acquire because they were too small to be observed. But Democritus could point out compelling, indirect evidence.

Consider a gold ring, for example. Over the years, the ring begins to wear down. Something is being lost. Every day some tiny bits of matter have been worn off the ring. Hence, although atoms are invisible, their existence can be measured indirectly.

Even today most of our advanced science is done indirectly. We know the composition of the sun, the detailed structure of DNA, and the age of the universe, all due to measurements of this kind. We know all this, even though we have never visited the stars or entered a DNA molecule or witnessed the Big Bang. This distinction between direct and indirect evidence will become essential when we discuss attempts to prove a unified field theory.

A second approach was pioneered by the great mathematician Pythagoras.

Pythagoras had the insight to apply a mathematical description to worldly phenomena like music. According to legend, he noticed similarities between the sound of plucking a lyre string and the resonances made by hammering a metal bar. He found that they created musical frequencies that vibrated with certain ratios. So something as aesthetically pleasing as music has its origin in the mathematics of resonances. This, he thought, might show that the diversity of the objects we see around us must obey these same mathematical rules.

So at least two great theories of our world emerged from ancient Greece: the idea that everything consists of invisible, indestructible atoms and that the diversity of nature can be described by the mathematics of vibrations.

Unfortunately, with the collapse of classical civilization, these philosophical discussions and debates were lost. The concept that there could be a paradigm explaining the universe was forgotten for almost a thousand years. Darkness spread over the Western world, and scientific inquiry was largely replaced by belief in superstition, magic, and sorcery.

REBIRTH DURING THE RENAISSANCE

In the seventeenth century, a few great scientists rose to challenge the established order and investigate the nature of the universe, but they faced stiff opposition and persecution. Johannes Kepler, who was one of the first to apply mathematics to the motion of the planets, was an imperial adviser to Emperor Rudolf II and perhaps escaped persecution by piously including religious elements in his scientific work.

The former monk Giordano Bruno was not so lucky. In 1600, he was tried and sentenced to death for heresy. He was gagged, paraded naked in the streets of Rome, and finally burned at the stake. His chief crime? Declaring that life may exist on planets circling other stars.

The great Galileo, the father of experimental science, almost met the same fate. But unlike Bruno, Galileo recanted his theories on pain of death. Nonetheless, he left a lasting legacy with his telescope, perhaps the most revolutionary and seditious invention in all of science. With a telescope, you could see with your own eyes that the moon was pockmarked with craters; that Venus had phases consistent with its orbiting the sun; that Jupiter had moons, all of which were heretical ideas.

Sadly, he was placed under house arrest, isolated from visitors, and eventually went blind. (It was said because he once looked directly at the sun with his telescope.) Galileo died a broken man. But the very year that he died, a baby was born in England who would grow up to complete Galileo's and Kepler's unfinished theories, giving us a unified theory of the heavens.

NEWTON'S THEORY OF FORCES

Isaac Newton is perhaps the greatest scientist who ever lived. In a world obsessed with witchcraft and sorcery, he dared to write down the universal laws of the heavens and apply a new mathematics he invented to study forces, called the calculus. As physicist Steven Weinberg has written, "It is with Isaac Newton that the modern dream of a final theory really begins." In its time, it was considered to be the theory of everything—that is, the theory that described all motion.

It all began when he was twenty-three years old. Cambridge University was closed because of the black plague. One day in 1666, while walking around his country estate, he saw an apple fall. Then he asked himself a question that would alter the course of human history.

If an apple falls, then does the moon also fall?

Before Newton, the church taught that there were two kinds of laws. The first were the laws found on Earth, which were corrupted by the sin of mortals. The second were the pure, perfect, and harmonious laws of the heavens.

The essence of Newton's idea was to propose a unified theory that encompassed the heavens and the Earth.

In his notebook, he drew a fateful picture (see figure 1).

If a cannonball is fired from a mountaintop, it goes a certain distance before hitting the ground. But if you fire the cannonball at increasing velocities, it travels farther and farther before coming back to Earth, until it eventually com-

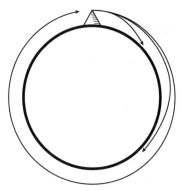

Figure 1. One can fire a cannonball with increasing energy, so that it eventually goes completely around the Earth and returns to its starting point. Newton then said that this explains the orbit of the moon, thereby unifying the physical laws found on Earth with the laws of heavenly bodies.

pletely circles the Earth and returns to the mountaintop. He concluded that the natural law that governs apples and cannonballs, gravity, also grips the moon in its orbit around the Earth. Terrestrial and heavenly physics were the same.

The way he accomplished this was to introduce the concept of forces. Objects moved because they were pulled or pushed by forces that were universal and could be measured precisely and mathematically. (Previously, some theologians thought that objects moved because of desires, so that objects fell because they yearned to be united with the Earth.)

Thus, Newton introduced the key concept of unification.

But Newton was a notoriously private man and kept much of his work a secret. He had few friends, was incapable of small talk, and was often immersed in bitter priority battles with other scientists about his discoveries.

In 1682, a sensational event happened that changed the course of history. A blazing comet sailed over London. Everyone, from kings and queens to beggars, was buzzing with the news. Where did it come from? Where was it going? What did it portend?

One man who took an interest in this comet was astronomer Edmond Halley. He took a trip to Cambridge to meet the famous Isaac Newton, already well-known for his theory of light. (By shining sunlight through a glass prism, Newton showed that white light separated into all the colors of the rainbow, thereby demonstrating that white light is actually a composite color. He also invented a new type of telescope that used reflecting mirrors rather than lenses.) When Halley asked Newton about the comet that everyone was talking about, he was shocked to hear that Newton could show that comets moved in ellipses around the sun and that he could even predict their trajectory using his own theory of gravity. In fact, he was tracking them with the telescope he invented, and they moved just as he predicted.

Halley was stunned.

He immediately realized that he was witnessing a landmark in science and volunteered to pay for the printing costs of what would eventually become one of the greatest masterpieces in all science, *Mathematical Principles of Natural Philosophy,* or simply *Principia.*

Furthermore, Halley, realizing that Newton was predicting that comets could return at regular intervals, calculated that the comet of 1682 would return in 1758. (Halley's comet sailed over Europe on Christmas Day, 1758, as predicted, helping to seal Newton's and Halley's reputations posthumously.)

Newton's theory of motion and gravitation stands as one of the greatest achievements of the human mind, a single principle unifying the known laws of motion. Alexander Pope wrote:

Nature and Nature's laws lay hid in night:
God said, Let Newton be!
And all was light.

Even today, it is the laws of Newton that allow NASA engineers to guide our space probes across the solar system.

WHAT IS SYMMETRY?

Newton's law of gravity is also noteworthy because it possesses a symmetry, so that the equation remains the same if we make a rotation. Imagine a sphere surrounding the Earth. The force of gravity is identical at every point on it. In fact, that is why the Earth is spherical, rather than another shape: because gravity compressed the Earth uniformly. That is why we never see cubical stars or pyramidal planets. (Small asteroids are often shaped irregularly, because the gravitational force on an asteroid is too small to compress it evenly.)

The concept of symmetry is simple, elegant, and intuitive. Moreover, throughout this book, we will see that symmetry is not just frivolous window dressing to a theory, but in fact is an essential feature that indicates some deep, underlying physical principle about the universe.

But what do we mean when we say an equation is symmetrical?

An object is symmetrical if, after you rearrange its parts, it is left the same, or invariant. For example, a sphere is symmetrical because it remains the same after you rotate it. But how can we express this mathematically?

Think of the Earth revolving around the sun (see figure 2). The radius of the Earth's orbit is given by R, which remains the same as the Earth moves in its orbit (actually, the Earth's orbit is elliptical so R varies slightly, but that's not important for this example). The coordinates of the Earth's orbit are given by X and Y. As the Earth moves in its orbit, X and Y continually change, but R is invariant—that is it doesn't change.

So the equations of Newton maintain this symmetry, meaning that the gravity between the Earth and the sun remains the same as the Earth orbits the sun. As our frame of reference changes, the laws stay constant. No matter what orientation we take looking at a problem, the rules are unchanging, and the results come out the same.

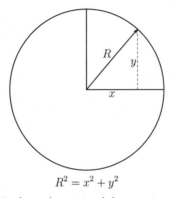

$$R^2 = x^2 + y^2$$

Figure 2. If the Earth revolves around the sun, its radius *R* remains the same. The coordinates *X* and *Y* of the Earth continually change as it orbits, but *R* is an invariant. By the Pythagorean theorem, we know that $X^2 + Y^2 = R^2$. So Newton's equation has a symmetry when expressed either in terms of *R* (because *R* is an invariant) or *X* and *Y* (via the Pythagorean theorem).

We will encounter this concept of symmetry over and over when we discuss the unified field theory. In fact, we will see that symmetry is one of our most powerful tools in unifying all the forces of nature.

CONFIRMATION OF NEWTON'S LAWS

Over the centuries, numerous confirmations of Newton's laws have been found, and they had a tremendous impact on science and also society. In the nineteenth century, astronomers noticed a strange anomaly in the heavens. The planet Uranus was deviating from the predictions of Newton's laws. Its orbit was not a perfect ellipse, but wobbled a bit. Either Newton's laws were flawed, or there was a planet that was not yet discovered whose gravity was tugging on the orbit of Uranus. Faith in Newton's laws was so great that physicists like Urbain Le Verrier tediously calculated where this mystery planet might lie. In 1846, on the very first try, astronomers found this planet to within one degree of where it was predicted to be. The new planet was dubbed Neptune. This was a tour de force for Newton's laws, and the first time in history that pure mathematics was used to detect the presence of a major celestial body.

As we mentioned earlier, every time scientists decoded one of the four fundamental forces of the universe, it not only revealed the secrets of nature, it also revolutionized society itself. Newton's laws not only unlocked the secret of the planets and comets, they also laid the foundation of the laws of mechanics, which we use today to design skyscrapers, engines, jet planes, trains, bridges, submarines, and rockets. For example, in the 1800s physicists applied Newton's laws to explain the nature of heat. At the time, scientists speculated that heat was some form of liquid that spread through a substance. But further investigation showed that heat was actually molecules in motion, resembling tiny steel balls constantly colliding with one another.

Newton's laws allowed us to calculate precisely how two steel balls bounced off each other. Then, by adding trillions upon trillions of molecules, one could calculate the precise properties of heat. (For example, when a gas in a chamber is heated, it expands according to Newton's laws since the heat increases the velocity of the molecules inside the chamber.)

Engineers could then use these calculations to perfect the steam engine. They could calculate how much coal was needed to turn water into steam, which could then be used to push gears, pistons, wheels, and levers to power machines. With the coming of the steam engine in the 1800s, the energy available to a worker skyrocketed to hundreds of horsepower. Suddenly, steel rails were connecting distant parts of the world and vastly increasing the flow of goods, knowledge, and people.

Before the Industrial Revolution, goods were made by tiny, exclusive guilds of skilled craftsmen who toiled to create even the simplest household items. They also jealously guarded the secrets of their handicraft. Hence, goods were often scarce and expensive. With the coming of the steam engine and the powerful machines it made possible, goods could be stamped out at a fraction of the original cost, vastly increasing the collective wealth of nations and raising our standard of living.

When I teach Newton's laws to promising engineering students, I try to emphasize that these laws are not just dry, boring equations, but they have changed the course of modern civilization, creating the wealth and prosperity we see all around us. We sometimes even show our students the catastrophic collapse of the Tacoma Narrows Bridge in Washington State in 1940, recorded on film, as a stunning example of what happens when we misapply Newton's laws.

Newton's laws, based on unifying the physics of the heavens with the physics of the Earth, helped to usher in the first great revolution in technology.

MYSTERY OF ELECTRICITY AND MAGNETISM

It would take another two hundred years for the next big breakthrough, which came from the study of electricity and magnetism.

The ancients had known that magnetism could be tamed; the invention of the compass by the Chinese harnessed the power of magnetism and helped launch an age of discovery. But the ancients feared the power of electricity. Lightning bolts were thought to express the anger of the gods.

The man who finally laid the foundation for this field was Michael Faraday, a poor but industrious youth who lacked any formal education. As a child, he managed to get a job working as an assistant at the Royal Institution in London. Normally, someone of his low social standing would forever sweep the floor, wash bottles, and hide in the shadows. But this young man was so tireless and inquisitive that his supervisors allowed him to perform experiments.

Faraday would go on to make some of the greatest discoveries in electricity and magnetism. He showed that if you take a magnet and move it inside a hoop of wire, then electricity is generated in the wire. This was an amazing and important observation, since the relationship between electricity and magnetism was then totally unknown. One could also show the reverse, that a moving electric field can create a magnetic one.

It gradually dawned on Faraday that these two phenomena were actually two sides of the same coin. This simple observation would help to open up the electric age, in which giant hydroelectric dams would light up entire cities. (In a hydroelectric dam, the river pushes against a wheel that spins a magnet that then pushes electrons inside a wire that sends the electricity to the sockets in your home. The opposite effect, turning electric fields into magnetic ones, is the reason why your vacuum cleaner works. Electricity from the wall socket causes a magnet to spin, which drives a pump creating suction and causes the rollers of the vacuum cleaner to spin as well.)

But because Faraday had no formal education, he did not have the command of the mathematics that would allow him to describe his remarkable discoveries. Instead, he filled up notebooks with strange diagrams showing lines of force that look like the web of lines iron filings make when surrounding a magnet. He also invented the concept of a field, one of the most important concepts in all of physics. A field consists of these lines of force spread throughout space. Magnetic lines surround every magnet, and the magnetic field of the Earth emanates from the north pole, spreads through space, and then returns to the south pole. Even Newton's theory of gravity can be expressed in terms of fields, so that the Earth moves around the sun because it moves in the sun's gravitational field.

Faraday's discovery helped to explain the origin of the magnetic field surrounding the Earth. Since the Earth spins, the electric charges inside the Earth also spin. This constant motion moving inside the Earth is responsible for the magnetic field. (But this still left open a mystery: Where does the magnetic field of a bar magnet come from, since there is nothing moving or spinning in it? We will return to this mystery later.) Today, all the known forces of the universe are expressed in the language of fields first introduced by Faraday.

Given Faraday's immense contribution to initiating the electric age, physicist Ernest Rutherford declared him the "greatest scientific discoverer of all time."

Faraday was also unusual at least for his time because he loved to engage the public, and even children, in his discoveries. He was famous for his Christmas Lectures, where he would invite everyone to the Royal Institution in London to witness dazzling displays of electrical wizardry. He would enter a large room whose walls were covered with metal foil (which today is called a Faraday cage), and then electrify it. Although the metal was clearly electrified, he was totally safe because the electric field spread out over the entire surface of the room, so the electric field inside remained zero. Today, this effect is commonly used to shield microwave ovens and delicate equipment from stray electric fields, or to protect jet planes, which are often struck by lightning bolts. (For a Science Channel program I once hosted, I went inside a Faraday cage at the Boston Museum of Science. Huge bolts of electricity, up to two million volts, bombarded the cage, filling the auditorium with a loud crackling sound. But I did not feel a thing.)

MAXWELL'S EQUATIONS

Newton had shown that objects move because they were pushed by forces, which could be described by calculus. Faraday showed that electricity moved because it was pushed by a field. But the study of fields required a new branch of mathematics, which was eventually codified by Cambridge mathematician James Clerk Maxwell and called vector calculus. So in the same way that Kepler and Galileo laid the foundation for Newtonian physics, Faraday paved the way for Maxwell's equations.

Maxwell was a virtuoso in mathematics who made astonishing breakthroughs in physics. He realized that the behavior of electricity and magnetism, as discovered by Faraday and others, could be summarized in precise mathematical language. One law stated that a moving magnetic field could create an electric field. Another law stated the opposite, that a moving electric field could create a magnetic field.

Then Maxwell had an idea for the ages. What if a changing electric field created a magnetic one that then created another electric field that then created another magnetic field, etc.? He had the brilliant insight that the end product of this rapid back-and-forth motion would be a moving wave, where electric and magnetic fields were constantly turning into each other. This infinite sequence of transformations has a life of its own, creating a moving wave of vibrating electric and magnetic fields.

Using vector calculus, he calculated the speed of this moving wave, and he found it to be 310,740 kilometers per second. He was shocked beyond belief. To within experimental error, this speed was remarkably close to the speed of light (which is now known to be 299,792 kilometers per second). He then made the next bold step to claim that this *was* light! Light is an electromagnetic wave.

Maxwell then wrote prophetically, "We can scarcely avoid the inference that light consists in the transverse undulations of the same medium which is the cause of electric and magnetic phenomena."

Today, every physics student and electrical engineer has to memorize Maxwell's equations. They are the basis for TVs, lasers, dynamos, generators, etc.

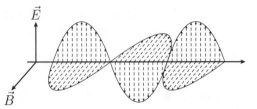

Figure 3. Electrical and magnetic fields are two sides of the same coin. Oscillating electric and magnetic fields turn into each other and move like a wave. Light is one manifestation of an electromagnetic wave.

Faraday and Maxwell unified electricity and magnetism. And the key to unification is symmetry. Maxwell's equations contain the symmetry called duality. If the electric within a light beam is represented by E and the magnetic field by B, then the equations for electricity and magnetism remain the same when we switch E and B. This duality implies that electricity and magnetism are two manifestations of the same force. So the symmetry between E and B allows us to unify electricity and magnetism, thereby creating one of the greatest breakthroughs of the nineteenth century.

Physicists were fascinated by this discovery. The Berlin Prize was offered to anyone who could actually reproduce these Maxwell waves in the laboratory. In 1886, physicist Heinrich Hertz performed the historic test.

First, Hertz created an electric spark in one corner of his laboratory. Several feet away, he had a coil of wire. Hertz showed that by turning on the spark, he could generate an electrical current in the coil, thereby proving that a new, mysterious wave traveled wirelessly from one place to another. This heralded the creation of a new kind of phenomenon, called radio. In 1894, Guglielmo Marconi introduced this new form of communication to the public. He showed that you could send wireless messages across the Atlantic Ocean at the speed of light.

With the introduction of radio, we now had a superfast, convenient, and wireless way of communicating over long distances. Historically, the lack of a fast and reliable communication system was one of the great obstacles to the march of history. (In 490 BCE, after the Battle of Marathon between the Greeks and the Persians, a poor runner was ordered to spread the news of the Greek victory as fast as he could. Bravely, he ran 26 miles to Athens after previously running 147 miles to Sparta, and then, according to legend, dropped dead of sheer exhaustion. His heroism, in the age before telecommunication, is now celebrated in the modern marathon.)

Today, we take for granted that we can send messages and information effortlessly across the globe, utilizing the fact that energy can be transformed in many ways. For example, when speaking on a cell phone, the energy of the sound of your voice converts to mechanical energy in a vibrating diaphragm. The diaphragm is attached to a magnet that relies on the interchangeability of electricity and magnetism to create an electrical impulse, the kind that can be transported and read by a computer. This electrical impulse is then translated into electromagnetic waves that are picked up by a nearby microwave tower. There, the message is amplified and sent across the globe.

But Maxwell's equations not only gave us nearly instantaneous communication via radio, cell phone, and fiber-optic cables, they also opened up the entire electromagnetic spectrum, of which visible light and radio were just two members. In the 1660s, Newton had shown that white light, when sent through a prism, can be broken up into the colors of the rainbow. In 1800, William Herschel had asked himself a simple question: What lies beyond the colors of the rainbow, which extend from red to violet? He took a prism, which created a rainbow in his lab, and placed a thermometer below the color red, where there was no color at all. Much to his surprise, the temperature of this blank area began to rise. In other words, there was a "color" below red that was invisible to the naked eye but contained energy. It was called infrared light.

Today, we realize that there is an entire spectrum of electromagnetic radiation, most of which is invisible, and each has a distinct wavelength. The wavelength of radio and TV, for example, is longer than that of visible light. The wavelength of the colors of the rainbow, in turn, is longer than that of ultraviolet and X-rays.

This also meant that the reality we see all around us is only the tiniest sliver of the complete EM spectrum, the smallest approximation of a much larger universe of EM

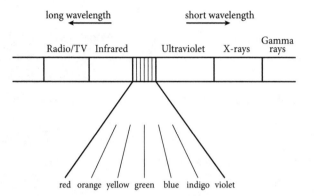

Figure 4. Most of the "colors" of the EM spectrum, extending from radio to gamma rays, are invisible to our eyes. Our eyes can only see the tiniest sliver of the entire EM spectrum, due to the size of the cells in our retinas.

colors. Some animals can see more than we can. For example, bees can see ultraviolet light, which is invisible to us but essential for them to find the sun even on a cloudy day. Since flowers evolved their gorgeous colors in order to attract insects like bees to pollinate them, this means that flowers are often even more spectacular when viewed using UV light.

When I was a child and read about this, I wondered why we could only see the smallest piece of the EM spectrum. What a waste, I thought. But the reason, I now realize, is that the wavelength of an EM wave is roughly the size of the antenna that produces it. Hence, your cell phone is only a few inches in size because that is the size of the antenna, which is about the wavelength of the EM waves being broadcasted. Similarly, the size of a cell in your retina is about the size of the wavelength of colors you can see. Hence we can see only colors whose wavelengths are the size of our cells. All the other colors of the EM spectrum are invisible because they are either too big or too small to be detected by our retinal cells. So if the cells of our eyes were the size of a house, we might be able to see all the radio and microwave radiation swirling around us.

Similarly, if the cells of our eyes were the size of atoms, we might be able to see X-rays.

Yet another application of Maxwell's equation is the way in which EM energy can power the whole planet. Although oil and coal have to be shipped by boat and train over vast distances, electrical energy can be sent over wires with the flick of a switch, electrifying entire cities.

This, in turn, led to a famous controversy between two giants of the electric age, Thomas Edison and Nikola Tesla. Edison was the genius behind many electrical inventions including the light bulb, motion pictures, the phonograph, the ticker tape, and hundreds of other marvels. He was also the first to wire a street with electricity, in this case Pearl Street in downtown Manhattan.

This created the second great revolution in technology, the electric age.

Edison assumed that direct current, or DC (which always moves in the same direction and never varies in voltage), would be the best way to transmit electricity. Instead of

DC power, Tesla, who used to work for Edison and helped lay the groundwork for the telecommunication network of today, advocated AC power (alternating current, so electricity reverses direction about sixty times a second). This resulted in the celebrated battle of the currents, with giant corporations investing millions in rival technologies, with General Electric backing Edison and Westinghouse backing Tesla. The future of the electric revolution would hinge on who won this conflict, Edison's DC or Tesla's AC.

Although Edison was the mastermind behind electricity and one of the architects of the modern world, he did not fully understand Maxwell's equations. This would be a very costly mistake. In fact, he thumbed his nose at scientists who knew too much mathematics. (In a famous story, he would often ask scientists looking for a job to calculate the volume of a light bulb. He would smile as these scientists tried to use advanced mathematics to tediously calculate the shape of the light bulb and then its volume. Afterward, Edison would simply pour water into an empty light bulb and then pour it into a graduated beaker.)

Engineers knew that wires strung over many miles lost a significant amount of energy if they carried low voltages, as advocated by Edison. So Tesla's high-voltage power lines were economically preferred, but high-voltages cables were too dangerous to be introduced into your living room. The trick was to use efficient high-voltage cables from the power plant to your city, and then somehow transform the high voltage to low voltage just before it entered your living room. The key was to use transformers.

As we recall, Maxwell showed that a moving magnetic field created an electric current, and vice versa. This allows you to create a transformer that can rapidly change the voltage in a wire. For example, the voltage of the electrical cables from a power station may carry thousands of volts. But the transformer located just outside your house can reduce the voltage to 110 volts, which easily powers your microwave oven and refrigerator.

If these fields are static and do not change, then they cannot be converted into each other. Because it is constantly changing, AC electricity can easily be converted into magnetic fields that are then converted back into electric fields, but at a lower voltage, meaning that AC current can easily change voltage using transformers; but DC current (because it is constant in voltage and not alternating) cannot.

In the end, Edison lost the battle and the considerable funds he invested in DC technology. That is the price of ignoring Maxwell's equations.

THE END OF SCIENCE?

In addition to explaining the mysteries of nature and bringing in a new era of economic prosperity, a combination of

Newton's and Maxwell's equations gave us a very convincing theory of everything. Or at least everything then known.

By 1900 prominent scientists were proclaiming the "end of science." Thus, the turn of the last century was a heady time to be alive. Everything that could be discovered had already been discovered, or so it seemed.

Physicists at that time did not realize that the two great pillars of science, Newton's and Maxwell's equations, were actually incompatible. They contradicted each other.

One of these two great pillars had to fall. And a sixteen-year-old boy held the key. That boy would be born the very year that Maxwell died, 1879.

2

EINSTEIN'S QUEST FOR UNIFICATION

As a teenager, Einstein asked himself a question that would alter the course of the twentieth century. He asked himself:

Can you outrace a light beam?

Years later, he would write that this simple question contained the key to his theory of relativity.

Earlier, he had read a children's book, Aaron David Bernstein's *Popular Books on Natural Science,* that asks you to imagine racing alongside a telegraph wire. Instead, Einstein envisioned running along a light beam, which should look frozen. Racing neck and neck alongside the beam, the light waves should be stationary, he thought, as Newton might have predicted.

But even as a sixteen-year-old boy, Einstein realized that no one had ever seen a frozen light beam before. Something was missing. He would ponder this question for the next ten years.

Unfortunately, many people considered him to be a failure. Although he was a brilliant student, his professors hated his freewheeling, bohemian lifestyle. Because he already knew most of the material, he often cut classes, so his professors wrote unflattering letters of recommendation; and every time he applied for a job he was turned down. Unemployed and desperate, he took a job tutoring (from which he got fired for arguing with his employer). He once considered selling insurance to support his girlfriend and child. (Can you imagine opening your door one day and seeing Einstein trying to sell you insurance?) Unemployable, he considered himself to be a drain on his family. In one letter, he wrote despondently, "I am nothing but a burden to my relatives. . . . It would surely be better if I did not live at all."

He finally managed to get a job as a clerk, third class, at the patent office in Bern. It was humiliating but actually a blessing in disguise. In the quiet of the patent office Einstein could return to the old question that had haunted him since he was a child. From there, he would launch a revolution that turned physics and the world upside down.

As a student at the famed École Polytechnique in Switzerland, he had come across Maxwell's equations for light. He asked himself, what happens to Maxwell's equations if you travel at the speed of light? Remarkably, no one had ever asked that question before. Using Maxwell's theory, Einstein calculated the speed of a light beam in a moving object, such as a train. He expected that the speed of the light beam, as seen by a stationary outside observer, would simply be its usual speed plus the speed of the train. According to Newtonian mechanics, velocities can simply add. For example, if you throw a baseball while riding on a train, a stationary observer would say that the speed of the ball is just the speed of the train plus the speed of the ball relative to the train. Likewise, velocities can also subtract. So, if you traveled neck and neck alongside a light beam, it should look stationary.

To his shock, he found that the light beam was not frozen at all but sped away at the same velocity. But this was impossible, he thought. According to Newton you can always catch up with anything if you move fast enough. That was common sense. But Maxwell's equations said that you could never catch up to light, which always moved at the same velocity, no matter how fast you traveled.

To Einstein, this insight was monumental. Either Newton or Maxwell was correct. The other had to be wrong. But how could it be that you could never catch up to light? At the patent office, he had plenty of time to ponder this question. One day, in the spring of 1905, it struck him while riding the train in Bern. "A storm broke loose in my mind," he would recall.

His brilliant insight was that since the speed of light is measured by clocks and metersticks, and since the speed of light is constant no matter how fast you move, space and time must be distorted in order to keep the speed of light constant!

It meant that if you are on a fast-moving spaceship, then clocks inside the ship beat slower than clocks on the Earth. *Time slows down the faster you move*—this phenomenon is described by Einstein's special relativity. So the question What time is it? depends on how fast you have been moving. If the rocket ship is traveling near the speed of light, and we observe it from the ground using a telescope, everyone in the rocket ship seems to move in slow motion. Also, everything in the ship seems to be compressed. Finally, everything in the rocket ship is heavier. Surprisingly, to someone in the rocket ship, everything appears normal.

Einstein would later recall, "I owe more to Maxwell than to anyone." Today, this experiment can be done routinely. If you place an atomic clock on an airplane, and compare it with a clock on the Earth, you can see that the clock on the airplane has slowed down (by a small factor, one part in a trillion).

But if space and time can vary, then everything you can measure must also vary, including matter and energy. And the faster you move, the heavier you become. But where does the extra mass come from? It comes from the energy of motion. This means that some of the energy of motion is turned into mass.

The precise relationship between matter and energy was $E = mc^2$. This equation, as we shall see, answered one of the deepest questions in all of science: Why does the sun shine? The sun shines because when you compress hydrogen atoms at great temperatures, some of the mass of the hydrogen gets converted to energy.

The key to understanding the universe is unification. For relativity, it was the unification of space and time and matter and energy. But how is this unification accomplished?

SYMMETRY AND BEAUTY

To poets and artists, beauty is an ethereal, aesthetic quality that evokes great emotion and passion.

To a physicist, beauty is symmetry. Equations are beautiful because they have a symmetry—that is, if you rearrange or reshuffle the components, the equation remains the same. It is invariant under this transformation. Think of a kaleidoscope. It takes a random jumble of colored shapes and, with mirrors, makes numerous copies and then arranges these images symmetrically in a circle. So something that is chaotic suddenly becomes ordered and beautiful because of symmetry.

Similarly, a snowflake is beautiful because, if we rotate it by 60 degrees, it remains the same. A sphere has even more symmetry. You can rotate it by any amount around its center, and the sphere looks identical. To a physicist, an equation is beautiful if we rearrange its various particles and components inside the equation and find the result does not change—in other words, if we find it has symmetry among its parts. The mathematician G. H. Hardy once wrote, "A mathematician's patterns, like the painter's or the poet's, must be *beautiful;* the ideas, like the colors or the words, must fit together in a harmonious way. Beauty is the first test; there is no permanent place in the world for ugly mathematics." And that beauty is symmetry.

We saw earlier that if you take Newton's gravitational force for the Earth going around the sun, the radius of the Earth's orbit is constant. The coordinates X and Y change, but R does not. This can also be generalized to three dimensions.

Imagine sitting on the surface of the Earth, where your location is given by three dimensions: X, Y, and Z are your coordinates (see figure 5). As you travel anywhere along the surface of the Earth, the radius of the Earth, R, remains the same, where $R^2 = X^2 + Y^2 + Z^2$. This is a three-dimensional version of the Pythagorean theorem.*

---

* To see this, let us take $Z = 0$. Then the sphere reduces down to a circle in the X and Y plane, just as before. We saw that as you move around this circle, we have $X^2 + Y^2 = R^2$. Now, let us gradually increase Z. The circle gets smaller as we rise in the Z direction. (The circle corresponds to the lines of equal latitude on a globe.) R remains the same, but the equation for the small circle becomes $X^2 + Y^2 + Z^2 = R^2$, for a fixed

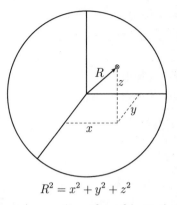

$$R^2 = x^2 + y^2 + z^2$$

Figure 5. As you wander over the surface of the Earth, the radius, *R*, of Earth is a constant, an invariant, although your coordinates *X, Y,* and *Z* constantly change into one another. So the three-dimensional Pythagorean theorem is the mathematical expression of this symmetry.

Now, if we take Einstein's equations and then rotate space into time and time into space, the equations remain the same. This means that the three dimensions of space are now joined with the dimension of time, *T,* which becomes the fourth dimension. Einstein showed that the quantity $X^2 + Y^2 + Z^2 - T^2$ (with time expressed in certain units) remains the same, which is a modified version of the

---

value of *Z*. Now, if we let *Z* vary, we see that any point on the sphere has coordinates given by *X, Y,* and *Z,* such that the three-dimensional Pythagorean theorem holds. So in summary, the points on a sphere can all be described by the Pythagorean theorem in three dimensions, such that *R* remains the same, but *X, Y,* and *Z* all vary as you move around the sphere. Einstein's great insight was to generalize this to four dimensions, with the fourth dimension being time.

Pythagorean theorem in four dimensions. (Notice that the time coordinate has an additional minus sign. This means that although relativity is invariant under rotations in four dimensions, the time dimension is treated slightly differently from the other three spatial dimensions.) So Einstein's equations are symmetric in four dimensions.

Maxwell's equations were first written down around 1861, the year the American Civil War began. Earlier, we noted that they possess a symmetry that turns electric and magnetic fields into each other. But Maxwell's equations possess an additional hidden symmetry. If we alter Maxwell's equations in four dimensions by interchanging X, Y, Z, and T among themselves as Einstein did in the 1910s, they remain the same. This means that, if physicists had not been blinded by the success of Newtonian physics, then relativity might have been discovered during the Civil War!

GRAVITY AS CURVED SPACE

Although Einstein showed that space, time, matter, and energy were all part of a larger four-dimensional symmetry, there was one glaring gap in his equation: they said nothing about gravity and accelerations. He was not satisfied. He wanted to generalize his earlier theory, which he called special relativity, so that it included gravity and accelerated motions, creating a more powerful general theory of relativity.

His colleague physicist Max Planck, however, warned him of the difficulty of creating a theory that combined relativity and gravity. He said, "As an older friend, I must advise against it. For in the first place, you won't succeed, and even if you do, no one will believe you." But then he added, "If you are successful, you will be called the next Copernicus."

It was obvious to any physicist that Newton's theory of gravity and Einstein's theory were at odds. If the sun were to suddenly disappear without a trace, then Einstein claimed that it would take eight minutes for the Earth to feel the absence. Newton's famous equation for gravity does not mention the speed of light. Hence gravity travels instantaneously, violating relativity, so the Earth should immediately feel the effect of the missing sun.

Einstein had pondered the question of light for ten years, from when he was sixteen to twenty-six. He would spend the next ten years, until he was thirty-six, concentrating on the theory of gravity. The key to the whole puzzle occurred to him one day while leaning back on his chair, almost causing him to fall over. In that brief instant, he realized that if he had fallen over, he would be weightless. Then he realized that this might be the key to a theory of gravity. He would fondly recall that it was "the happiest thought of his life."

Galileo realized that if you fell off a building, you would momentarily be weightless, but only Einstein realized how to exploit this fact to reveal the secret of gravity. Imagine for a moment being in an elevator and the cable is cut. You would fall, but the floor falls at the same rate, so inside the elevator, you begin to float, as if there is no gravity (at least until the elevator hits the ground). Inside the elevator, gravity was precisely canceled by the acceleration of a falling elevator. This is called the equivalence principle, that acceleration in one frame is indistinguishable from gravity in another frame.

When our astronauts in space appear weightless on TV, it is not because gravity has disappeared from space. There is plenty of gravity throughout the entire solar system. The reason is because their rocket is falling at exactly the same rate as they are. Like Newton's imaginary cannonball shot from a mountaintop, they and their capsule are both in free fall around the Earth. Thus, inside the ship, it is an optical illusion that they are weightless, since everything, including your body and the ship itself, are falling at the same rate.

Einstein then applied this to a children's merry-go-round. According to relativity, the faster you move, the flatter you become because space compresses. As it spins, the outer rim of the ride moves faster than the interior. This means that, because of relativity's effect on space-time, the rim contracts more than the interior since the rim is moving faster. But as the merry-go-round approaches the speed of light, the floor is distorted. It is no longer just a flat disc. The rim has shrunk while the center remains the same, so the surface is curved like an upside-down bowl.

Now imagine trying to walk on the curved floor of the merry-go-round—you cannot walk in a straight line. At first you might think there is an invisible force that tries to throw you off because the surface is warped or curved. So someone on the merry-go-round says that there is a centrifugal force pushing everything off it. But to someone outside, there is no external force at all, just the curvature of the floor.

Einstein put it all together. The force that causes you to fall on a merry-go-round is actually caused by the warping of the merry-go-round. The centrifugal force you feel is equivalent to gravity—that is, it is a fictitious force created by being on an accelerating frame. *In other words, acceleration in one frame is identical to the effect of gravity in another, which is due to space being curved.*

Now replace the merry-go-round with the solar system. The Earth goes around the sun, so we Earthlings have the illusion that the sun exerts a force of attraction, called gravity, on the Earth. But to someone outside the solar system, they would not see a force at all; they would observe that the space around the Earth has curved, so that empty space is pushing the Earth so that it goes in a circle around the sun.

Einstein had the brilliant observation that gravitational

attraction was actually an illusion. Objects moved not because they are pulled by gravity or the centrifugal force but because they are pushed by the curvature of space around it. *That's worth repeating: gravity does not pull; space pushes.*

Shakespeare once said that all the world is a stage, and we are actors making our entrances and exits. This was the picture adopted by Newton. The world is static, and we move on this flat surface, obeying Newton's laws.

But Einstein abandoned this picture. The stage, he said, is curved and warped. If you walk on it, you cannot walk in a straight line. You are constantly being pushed because the floor beneath your feet is curved, and you stagger like a drunk.

Gravitational attraction is an illusion. For example, you might be sitting in a chair right now, reading this book. Normally, you would say that gravity is pulling you down into your chair, and that is why you don't fly off into space. But Einstein would say that you are sitting in your chair because the Earth's mass warps the space above your head, and this warping pushes you into your chair.

Imagine putting a heavy shot put on a large mattress. It sinks into the bed, causing it to warp. If you shoot a marble along the mattress, it moves in a curved line. In fact, it will circle the shot put. From a distance, an observer may say that there is an invisible force pulling on the marble, forcing it to orbit. But close up, you see that there is no invisible force at all. The marble does not move in a straight line

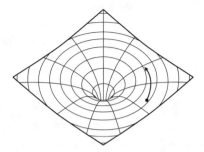

Figure 6. A heavy shot put on a mattress sinks into the fabric. A marble circles around the depression it creates. From a distance, it appears that a force from the shot put grabs the marble and forces it into an orbit. Actually, the marble is orbiting the shot put because the mattress is warped. In the same way, the sun's gravity warps the starlight from distant stars, which can be measured by telescopes during an eclipse of the sun.

because the mattress is curved, making the most direct path an ellipse.

Now replace the marble with the Earth, the shot put with the sun, and the mattress with space-time. Then we see that the Earth goes around the sun because the sun has warped the space around it, and the space Earth is traveling in is not flat.

Also, think of ants moving on a crumpled sheet of paper. They cannot move in a straight line. They might feel as if a force is continually tugging on them. But to us, looking down on the ants, we see that there is no force all. This is the insight of what Einstein called general relativity: space-time is warped by heavy masses, causing the illusion of gravitational force.

This means that general relativity is much more powerful and symmetrical than special relativity, since it describes gravity, which affects all things in space-time. Special relativity, on the other hand, only worked for objects moving smoothly in space and time in a straight line. But in our universe, almost everything is accelerating. From racing cars to helicopters to rockets, we see that they are all accelerating. General relativity works for accelerations that are continually changing at every point in space-time.

SOLAR ECLIPSE AND GRAVITY

Any theory, no matter how beautiful, must eventually confront experimental verification. So Einstein seized upon several possible experiments. The first was the erratic orbit of Mercury. When calculating its orbit, astronomers found a slight anomaly. Instead of moving in a perfect ellipse, as predicted by Newton's equations, it wobbled a bit, making a flowerlike pattern.

To protect Newton's laws, astronomers posited the existence of a new planet, called Vulcan, inside the orbit of Mercury. The gravity of Vulcan would tug on Mercury, causing the aberration. Earlier, we saw that this strategy allowed astronomers to discover the planet Neptune. But astronomers failed to find any observational evidence for Vulcan.

So when Einstein recalculated the perihelion of Mercury, the spot where it is closest to the sun, using his theory of gravity, he found a slight deviation from Newton's laws. He was ecstatic to find a perfect match with his own calculations. He found the difference from a perfect ellipse in its orbit to be 42.9 seconds of arc per century, well within the experimental result. He would recall fondly, "For some days, I was beyond myself with excitement. My boldest dreams have now come true."

He also realized that according to his theories light should be deflected by the sun.

Einstein realized that the sun's gravity would be powerful enough to bend the starlight of nearby stars. Since these stars could only be seen during a solar eclipse, Einstein proposed that an expedition be sent to witness the solar eclipse of 1919 to test his theory. (Astronomers would have to take two pictures of the night sky, one where the sun was absent and another during a solar eclipse. By comparing these two photographs, the position of the stars during the eclipse would have to move due to the sun's gravity.) He was certain his theory would be shown to be correct. When he was asked what he would think if the experiment disproved his theory, he said that God must have made a mistake. He was convinced he was correct, he wrote to his colleagues, because it had superb mathematical beauty and symmetry.

When this epic experiment was finally performed by astronomer Arthur Eddington, there was remarkable agreement between Einstein's prediction and the actual result.

(Today, the bending of starlight due to gravity is routinely used by astronomers. When starlight passes near a distant galaxy, light is bent, giving the appearance of a lens bending the light. These are called gravity lenses or Einstein lenses.)

Einstein would go on to win the Nobel Prize in 1921.

Soon, he became one of the most recognized figures on the planet, even more than most movie stars and politicians. (In 1933, he appeared with Charlie Chaplin at a movie premiere. When they were mobbed by autograph seekers, Einstein asked Chaplin, "What does all this mean?" Chaplin replied, "Nothing, absolutely nothing." Then he said, "They cheer me because everyone understands me. They cheer you because no one understands you.")

Of course, a theory that would overthrow 250 years of Newtonian physics would also be met with fierce criticism. One of the skeptics leading the charge was Columbia professor Charles Lane Poor. After reading about relativity, he fumed, "I feel as if I had been wandering with Alice in Wonderland and had tea with the Mad Hatter."

But Planck would always reassure Einstein. He would write, "A new scientific truth does not triumph by convincing its opponents and making them see the light, but rather because the opponents eventually die and a new generation grows up that is familiar with it."

Over the decades, there have been many challenges to relativity, but each time Einstein's theory has been verified. In fact, as we shall see in later chapters, Einstein's theory of relativity has reshaped the entire discipline of physics, revolutionizing our conception of the universe, its origin, and its evolution and changing the way we live.

One easy way to confirm Einstein's theory is to use the GPS system on your cell phone. The GPS system consists of thirty-one satellites orbiting the Earth. At any time, your cell phone can receive signals from three of them. Each of these three satellites is moving in a slightly different trajectory and angle. The computer in your cell phone then analyzes this data from the three satellites and triangulates your precise position.

The GPS system is so accurate that it has to take tiny corrections from both special and general relativity into account.

Since the satellites are moving at roughly 17,000 miles per hour, a clock in the GPS satellites beats slightly slower those than on Earth due to special relativity, which states that higher speeds result in slower time—the phenomenon demonstrated in Einstein's thought experiment of outracing a light beam. But since gravity is weaker the farther you move into outer space, time actually speeds up a bit due to general relativity, which states that space-time can be warped by gravitational pull—the weaker the gravitational pull, the faster time moves. This means that special and general relativity work in opposite directions, with special relativity causing the signals to slow down, while general relativity causes the signals to speed up. Your cell phone then factors in both competing effects and tells you precisely where you are located. So without special and general relativity working in tandem, you would be lost.

NEWTON AND EINSTEIN: POLAR OPPOSITES

Einstein was heralded as the next Newton, but Einstein and Newton were polar opposites in personality. Newton was a loner, reticent to the point of being antisocial. He had no lifelong friends and was incapable of everyday conversation.

Physicist Jeremy Bernstein once said, "Everyone who had any substantial contact with Einstein came away with an overwhelming sense of the nobility of the man. A descriptive term for him that recurs again and again is 'humanitarian'—a reference to the simple, lovable quality of his character."

But both Newton and Einstein shared certain key characteristics. The first was the ability to concentrate and focus tremendous mental energy. Newton could forget to eat or sleep for days when concentrating on a single problem. He would stop in the middle of a conversation and scribble on whatever was available, sometimes a napkin or the wall. Similarly, Einstein could focus on a problem for years, even decades. He even suffered a near breakdown while working on the general theory.

Another characteristic they shared was the ability to visualize a problem in pictures. Although Newton could have written *Principia* entirely in terms of algebraic symbols, instead he filled the masterpiece with geometric diagrams. To use calculus with abstract symbols is relatively easy; but deriving them from triangles and squares can only be done by a master. Similarly, Einstein's theory is filled with diagrams of trains, metersticks, and clocks.

SEARCH FOR A UNIFIED THEORY

In the end, Einstein created two major theories. The first was special relativity, which governed the properties of light beams and space-time. It introduced a symmetry based on rotations in four dimensions. The second was general relativity, where gravity is unveiled as the bending of space-time.

But after these two monumental achievements, he tried to reach for a third, even greater achievement. He wanted a theory that would unify all the forces of the universe into a single equation. He wanted to use the language of field theory to create an equation that could combine Maxwell's theory of electricity and magnetism with his own theory of gravity. He tried for decades to unify these two, and failed. (Michael Faraday was actually the first to propose a unification of gravity with electromagnetism. Faraday used to go to the London Bridge and drop magnets, hoping to find some measurable effect of gravity on the magnet. He found none.)

One reason why Einstein failed was that, in the 1920s, there was a huge hole in our understanding of the world. It would take advances in a new theory, the quantum theory, for physicists to realize that there was a missing piece of the puzzle: the nuclear force.

But Einstein, although he was one of the founders of the quantum theory, ironically would become the quantum's greatest adversary. He would unleash a barrage of criticisms against the quantum theory. Over the decades, the theory has met every experimental challenge and has given us a deluge of wondrous electrical appliances that fill up our lives and workplaces. However, as we shall see, his deep, subtle philosophical objections to it resonate even now.

3

RISE OF THE QUANTUM

hile Einstein was single-handedly creating this vast new theory based on space and time and matter and energy, a parallel development in physics was unraveling this age-old question: What is matter made of? This would lead to the next great theory of physics, the quantum theory.

After Newton had finished his theory of gravity, he performed numerous experiments in alchemy, trying to understand the nature of matter. His bouts of depression, it is theorized, were because of his experiments with mercury, a poison known to cause neurological symptoms. However, little was known about the fundamental properties of matter, and little was learned from the work of these early alchemists, who spent much of their time and energy trying to convert lead into gold.

It would take several centuries to slowly reveal the secrets of matter. By the 1800s, chemists began to find and isolate the basic elements of nature—elements that, in turn, could not be decomposed into anything simpler. While the stunning advances in physics were pioneered by mathematics, the breakthroughs in chemistry came mainly from tedious hours toiling in a laboratory.

In 1869, Dmitry Mendeleyev had a dream, in which all the elements of nature fell into a table. Upon awakening, he quickly began to arrange the known elements into a regular table, showing that there was a pattern to the elements. Out of the chaos of chemistry suddenly came order and predictability. The sixty or so known elements could be arranged into this simple table, but there were gaps, and Mendeleyev was able to predict the properties of these missing elements. When these elements were actually found in the laboratory, as predicted, it sealed the reputation of Mendeleyev.

But why were the elements arranged in such a regular pattern?

The next development occurred in 1898, when Marie and Pierre Curie isolated a new series of unstable elements, never before seen. Without any power source, radium glowed brightly in the laboratory, violating one of the cherished principles of physics, the conservation of energy (that energy can never be created or destroyed). The energy of these radium rays seemed to come from nowhere. Clearly, a new theory would be necessary.

Until then, chemists believed that the fundamental

ingredients of matter, the elements, were eternal, that elements like hydrogen or oxygen were stable for all time. But in their laboratories, chemists could see that elements like radium were decaying into other elements, releasing radiation in the process.

It was also possible to calculate how quickly these unstable elements were decaying, which could be measured in thousands or even billions of years. The Curies' discoveries helped settle a long-standing debate. Geologists, amazed at the glacial pace of rock formations, realized that the Earth must be billions of years old. But Lord Kelvin, one of the giants of classical Victorian physics, calculated that a molten Earth would cool down in a matter of a few million years. Who was right?

As it turns out, it was the geologists. Lord Kelvin did not understand that a new force of nature, the one being discovered by the Curies called the nuclear force, could add to the Earth's heat. Since radioactive decay could take place over billions of years, it meant that the Earth's core could be heated by the decay of uranium, thorium, and other radioactive elements. So the enormous power of shattering earthquakes, thundering volcanoes, and slow, grinding continental drift all originate from the nuclear force.

In 1910, Ernest Rutherford put a piece of glowing radium in a lead box with a minuscule hole. A tiny beam of radiation emerged from the hole, aimed at a thin sheet of gold. It was expected that the atoms of gold would absorb the radiation. To his shock, he found that the beam from the radium went right through the sheet, as if it weren't there.

This was an astonishing result: it meant that atoms were composed primarily of empty space. We sometimes demonstrate this to students. We put a piece of harmless uranium in their hand and a Geiger counter beneath it, which can detect radiation. Students are shocked to hear the Geiger counter clicking away because their body *is* hollow.

In the early 1900s, the standard picture of the atom was the raisin pie model—that is, the atom was like a pie of positive charge, with raisins of electrons sprinkled inside. Gradually, a radically new picture of the atom began to emerge. The atom was basically hollow, consisting of a swarm of electrons circling a tiny dense core, called the nucleus. Rutherford's experiment helped prove this because his radioactive beam would occasionally be deflected by the tightly packed particles in the nucleus. By analyzing the number, frequency, and angles of deflection, he was able to estimate the size of the nucleus of the atom. It was one hundred thousand times smaller than the atom itself.

Later, scientists determined that the nucleus was, in turn, made of even tinier subatomic particles: protons (which carried positive charge) and neutrons (which carried no charge). The entire Mendeleyev table, it seemed, could be created using only three subatomic particles: the electron, proton, and neutron. But what equation did these particles obey?

QUANTUM REVOLUTION

Meanwhile, a new theory was being born that could explain all these mysterious discoveries. This theory would eventually unleash a revolution that would challenge everything we knew about the universe. It was called quantum mechanics. But what is the quantum anyway, and why is it so important?

The quantum was born in 1900 when German physicist Max Planck asked himself a simple question: Why do objects glow when hot? When humans first harnessed fire thousands of years ago, they noticed that hot objects glow with certain colors. Pottery makers had known for centuries that, as objects reach thousands of degrees, they change color, going from red to yellow to blue. (You can see this for yourself by simply lighting a match or candle. At the very bottom, the flame is hottest, and its color might be bluish. It is yellowish in the center and coolest at the top, where the flame is reddish.)

But when physicists tried to calculate this effect (called blackbody radiation) by applying the work of Newton and Maxwell to atoms, they discovered a problem. (A blackbody is an object that perfectly absorbs all radiation that falls on it. It is called black because the color black absorbs all light.) According to Newton, as atoms get hotter, they vibrate more rapidly. And according to Maxwell, vibrating charges, in turn, can emit electromagnetic radiation in the form of light. But when they calculated the radiation emitted from hot, vibrating atoms, the result defied expectations. At low frequencies, this model fit the data quite well. But at high frequency, the energy of light should eventually become infinite, which was ridiculous. To a physicist, infinity is just a sign that the equations aren't working, that they don't understand what is happening.

Max Planck then posited an innocent hypothesis. He supposed that energy, instead of being continuous and smooth as in Newton's theory, actually occurred in discrete packets he called quanta. When he adjusted the energy of these packets, he found that he could reproduce precisely the energy that radiated from hot objects. The hotter the object, the higher the frequency of radiation, corresponding to different colors on the light spectrum.

This is why a flame changes from red to blue as the temperature increases. This is also how we know the temperature of the sun. When you first hear that the surface of the sun is about 5,000 degrees Celsius, you may wonder: How do we know that? No one has ever been to the sun with a thermometer. But we know the temperature of the sun because of the wavelength of light it is emitting.

Planck then calculated the size of these packets of light energy, or quanta, and measured them in terms of a small constant h, Planck's constant, which is 6.6×10^{-34} Joule-seconds. (This number was found by Planck by adjusting the energy of these packets by hand, until he could perfectly fit the data.)

If we let Planck's constant gradually go to zero, then all the equations of the quantum theory reduce to the equations of Newton. (This means that the bizarre behavior of subatomic particles, which often violate common sense, gradually reduces to the familiar Newtonian laws of motion as Planck's constant is manually set to zero.) That is why we rarely see quantum effects in daily life. To our senses, the world seems very Newtonian because Planck's constant is a very small number and only affects the universe on the subatomic level.

These small quantum effects are called *quantum corrections,* and physicists spend entire lifetimes trying to calculate them. In 1905, the same year that Einstein discovered special relativity, he applied the quantum theory to light and showed that light was not just a wave but acted like a packet of energy, or a particle, that was called the photon. So light apparently had two faces: a wave as predicted by Maxwell, and a particle or photon as predicted by Planck and Einstein. A new picture of light was now emerging. Light was made of photons, which are quanta, or particles, but each photon created fields surrounding it (the electric and magnetic fields). These fields, in turn, were shaped like waves and obeyed Maxwell's equations. So we now have a beautiful relationship between particles and the fields that surround it.

If light had two faces, both as a particle and as a wave, then did the electron also have this bizarre duality? This was the next logical step, and it would have the most profound effect, shaking the world of modern physics and civilization itself.

ELECTRON WAVES

Physicists, to their shock, then found that electrons, which were once considered to be hard, point-like particles, could also act like waves. To demonstrate, take two parallel sheets of paper, one behind the other. You drill two slits in the first sheet, and then fire an electron beam at it. You would normally expect to find two spots on the second sheet, where the electron beams hit. Either the electron beam goes through the first or the second slit. Not both. That's just common sense.

But when the experiment is actually done, the pattern of dots on the second sheet appears to be arranged in a band of vertical lines, which is a phenomenon that occurs when waves interfere with each other. (The next time you take a bath, gently splash the surface at two places in synchronization, and you will see this interference pattern emerge, resembling a network of spiderwebs.)

But this means, in some sense, that the electrons went through both slits simultaneously. This was the paradox: How can a point particle, the electron, interfere with itself,

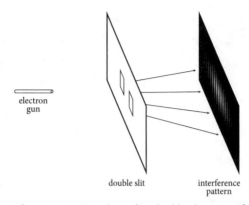

Figure 7. Electrons passing through a double slit act as if they are a wave—that is, they interfere with one another on the other side, as if they are moving through two slits simultaneously, which is impossible in Newtonian physics but is the basis of quantum mechanics.

as if it had traveled through two separate slits? In addition, other experiments on electrons showed they vanished and reappeared somewhere else, which is impossible in a Newtonian world. If Planck's constant were considerably larger, affecting things at a human scale, then the world would be a totally unrecognizable, bizarre place. Objects could disappear and reappear in a different location and could be two places at the same time.

As improbable as the quantum theory appeared to be, it began to have spectacular success. In 1925, Austrian physicist Erwin Schrödinger wrote down his celebrated equation that precisely described the motion of these particle waves. When applied to the hydrogen atom, with a single electron orbiting a proton, it gave remarkable agreement with experiment. The electron levels found in the Schrödinger atom exactly matched the experimental results. In fact, the entire Mendeleyev table could in principle be explained as a solution of the Schrödinger equation.

EXPLAINING THE PERIODIC TABLE

One of the spectacular achievements of quantum mechanics is its ability to explain the behaviors of the building blocks of matter, atoms and molecules. According to Schrödinger, the electron is a wave that surrounds the tiny nucleus. In figure 8, we see how only waves with certain discrete wavelengths can travel around the nucleus. Waves with an integral number of wavelengths fit nicely. But ones that do not have an integral number do not wrap fully around the nucleus. They are unstable and cannot form stable atoms. This means that electrons can only move in distinct shells.

As we go farther away from the nucleus, this basic pattern repeats itself; as the number of electrons increases, the outer ring moves farther away from the center. There are more electrons the farther you move. This in turn explains why the Mendeleyev table contains regular discrete levels that repeat themselves, with each level mimicking the behavior of the shell below it.

This effect is noticeable whenever you sing in the shower. Only certain discrete frequencies, or wavelengths, bounce

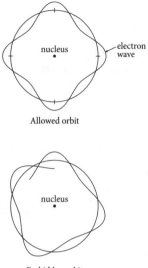

Allowed orbit

Forbidden orbit

Figure 8. Only electrons of a certain wavelength can fit inside an atom—that is, the orbit must be an integer multiple of the electron wavelength. This forces electron waves to form discrete shells around the nucleus. A detailed analysis of how electrons fill these shells can help to explain the Mendeleyev periodic table.

off the walls and are magnified, but others that don't fit are canceled, similar to the way electron waves circle the nucleus of an atom: only certain discrete frequencies work.

This breakthrough fundamentally changed the course of physics. One year, physicists were completely stumped when describing the atom. The next year, with Schrödinger's equation, they could calculate the properties inside the atom itself. I sometimes teach quantum mechanics to graduate students, and try to impress upon them the fact that everything around them, in a sense, can be expressed as a solution of his equation. I mention to them that not only can atoms be explained by it, but one can also explain the bonding of atoms to form molecules and therefore the chemicals from which our entire universe is composed.

No matter how powerful the Schrödinger equation was, however, it still had a limitation. It only worked for small velocities—that is, it was nonrelativistic. The Schrödinger equation said nothing about the speed of light, special relativity, and how electrons interact with light via Maxwell's equations. It also lacked the beautiful symmetry of Einstein's theory and was rather ugly and difficult to handle mathematically.

DIRAC THEORY OF THE ELECTRON

Then a twenty-two-year-old physicist, Paul Dirac, decided to write a wave equation that obeyed Einstein's special relativity by merging space and time. One of the things that was inelegant about the Schrödinger equation was that it treated space and time separately and hence calculations were often tedious and time-consuming. But Dirac's theory combined the two and had a four-dimensional symmetry, so it was also beautiful, compact, and elegant. All the ugly terms in the original Schrödinger equation collapsed into a simple four-dimensional equation.

(I remember when I was in high school, trying desperately to learn the Schrödinger equation, and struggling with all the ugly terms it contained. How could nature be so malicious, I thought, to create a wave equation that was so clumsy? Then one day, I stumbled upon the Dirac equation, which was beautiful and compact. I remember crying when I saw it.)

The Dirac equation was a spectacular success. We saw earlier that Faraday had shown that a changing electric field in a coil of wire produced a magnetic field. But where did the magnetic field come from in a bar magnet, without any moving charges? This seemed like a total mystery. But according to Dirac's equations, the electron was predicted to have a spin that created a magnetic field of its own. This property of the electron was built in from the very beginning in the mathematics. (This spin, however, is not the familiar spin we see around us—that is, as in a gyroscope— but is a mathematical term in the Dirac equation.) The magnetic field created by the spin matched precisely the field actually found surrounding electrons. This, in turn, helped to explain the origin of magnetism. So where does the magnetic field in a magnet come from? It comes from the spin of the electrons trapped inside the metal. Later, it was discovered that all subatomic particles have a spin. We will return to this important concept in a later chapter.

Even more important, the Dirac equation predicted an unexpected new form of matter, called antimatter. Antimatter obeys the same laws as ordinary matter, except it has the opposite charge. So the anti-electron, called the positron, has a positive, not a negative, charge. In principle, it may be possible to create anti-atoms, made of anti-electrons circling anti-protons and anti-neutrons. But when matter and antimatter collide, they explode in a burst of energy. (Antimatter will become a crucial ingredient of a theory of everything, since all particles in the final theory must have an antimatter counterpart.)

Previously, physicists considered symmetry to be an aesthetically pleasing but nonessential aspect of any theory. Now, physicists were staggered at the power of symmetry, that it could actually predict entirely new and unexpected physical phenomena (such as antimatter and electron spin). Physicists were beginning to understand that symmetry was an essential and inescapable feature of the universe at a fundamental level.

WHAT IS WAVING?

But there were still some nagging questions. If the electron had wavelike properties, then what was disturbing the medium in which the wave existed? What was waving? And how can it go through two different holes simultaneously? How can an electron be in two places at the same time?

The answer was startling and incredible, and split the physics community right down the middle. According to a paper by Max Born in 1926, *what was waving was the probability of finding an electron at that point.* In other words, you could not know for certain precisely where an electron was. All you could know was the probability of finding it. This was codified in Werner Heisenberg's celebrated uncertainty principle, which stated that you cannot know precisely the velocity and location of an electron. In other words, *electrons are particles, but the probability of finding the particle at any given location is given by a wave function.*

This idea was a bombshell. It meant that you could not accurately predict the future. You could only predict the odds that certain things will happen. But quantum theory's successes were undeniable. Einstein wrote that "the more successful the quantum theory becomes, the sillier it looks." Even Schrödinger, who had introduced the concept of the electron wave in the first place, rejected this probabilistic interpretation of his very own equations. Even today, there are arguments among physicists debating the philosophical implications of the wave theory. How can you be two places at the same time? Nobel laureate Richard Feynman once said, "I think I can safely say that nobody understands quantum mechanics."

Ever since Newton, physicists believed in something called determinism, the philosophy that all future events can be accurately predicted. The laws of nature determine the motion of all things in the universe, making them ordered and predictable. To Newton, the entire universe was a clock, beating in a precise predictable fashion. If you knew the location and velocity of all the particles in the universe, you could deduce all future events.

Predicting the future, of course, has always been an obsession of mortals. In *Macbeth,* Shakespeare wrote,

> *"If you can look into the seeds of time*
> *And say which grain will grow and which will not,*
> *Speak, then, to me."*

According to Newtonian physics, it is possible to predict which grain will grow and which will not. For several centuries, this view prevailed among physicists. So uncertainty was heresy, and shook modern physics to the core.

CLASH OF TITANS

On one side of this debate were Einstein and Schrödinger, who helped to start the quantum revolution in the first place. On the other side were Niels Bohr and Werner Heisenberg, creators of the new quantum theory. It culminated in the historic sixth Solvay Conference in 1930 in Brussels. It was to be a debate for the ages, when the giants of physics would go head-to-head to battle for the meaning of reality itself.

Paul Ehrenfest would write, "I will never forget the sight of the two opponents leaving the university club. Einstein, a majestic figure, walking calmly with a faint ironical smile, and Bohr trotting along by his side, extremely upset." Bohr could be heard muttering dejectedly to himself in the hallways, saying just one word, "Einstein . . . Einstein . . . Einstein."

Einstein led the charge, raising objection after objection to the quantum theory, trying to expose how absurd it was. But Bohr successfully countered each of Einstein's criticisms one by one. When Einstein kept repeating that God does not play dice with the universe, Bohr reportedly said, "Stop telling God what to do."

Princeton physicist John Wheeler said, "It was the greatest debate in intellectual history that I know about. In thirty years, I never heard of a debate between two greater men over a longer period of time on a deeper issue with deeper consequences for understanding this strange world of ours."

Historians agree for the most part that Bohr and the quantum rebels won the debate.

Still, Einstein was successful in exposing the cracks in the foundation of quantum mechanics. Einstein showed that it was a towering giant standing on philosophical feet of clay. These criticisms are heard even today, and they all center on a certain cat.

SCHRÖDINGER'S CAT

Schrödinger devised a simple thought experiment that exposed the essence of the problem. Place a cat in a sealed box. Put a piece of uranium in the box. When the uranium fires a subatomic particle, it triggers a Geiger counter that sets off a gun that fires a bullet at the cat. The question is: Is the cat dead or alive?

Since the firing of a uranium atom is a purely quantum event, it means that you have to describe the cat in terms of quantum mechanics. To Heisenberg, before you open the box, the cat exists as a mixture of different quantum states—that is, the cat is the sum of two waves. One wave describes a dead cat. The other wave describes a live cat. The cat is neither dead nor alive but a mixture of both. The only way to tell if the cat is dead or alive is to open the box and make an observation; then the wave function collapses into a dead or live cat. In other words, *observation (which requires consciousness) determines existence.*

To Einstein, all this was preposterous. It resembled the philosophy of Bishop Berkeley, who asked: If a tree falls in the forest and no one is there to hear it, does it make a sound? The solipsists would say no. But the quantum theory was even worse. It said that if there is a tree in the forest with no one around, the tree exists as the sum of many different states: for example, a burnt tree, a sapling, firewood, plywood, etc. Only when you look at the tree does its wave magically collapse into an ordinary tree.

When visitors would come to visit Einstein's house, he would ask them, "Does the moon exist because a mouse looks at it?" But no matter how much the quantum theory violated common sense, it had one thing going for it: it was experimentally correct. Predictions of the quantum theory have been tested to eleven decimal places, making it the most accurate theory of all time.

Einstein would admit, however, that the quantum theory contained at least *part* of the truth. In 1929, he even recommended Schrödinger and Heisenberg for the Nobel Prize in Physics.

Even today, there is no universal consensus among physicists concerning the cat problem. (The old Copenhagen interpretation of Niels Bohr, that the true cat emerges only because observation causes the wave of the cat to collapse, has fallen into disfavor, in part because with nanotechnology, we can now manipulate individual atoms and perform these experiments. What has become more popular is the multiverse, or many worlds, interpretation, where the universe splits in half, with one half containing a dead cat and the other containing a live cat.)

With the success of quantum theory, physicists in the 1930s then turned their sights to a new prize, answering the age-old question: Why does the sun shine?

ENERGY FROM THE SUN

Since time immemorial, the great religions of the world have exalted the sun, putting it at the very center of their mythology. The sun was one of the most powerful of the gods who ruled the heavens. To the Greeks, he was Helios, who grandly rode his blazing chariot across the sky every day, illuminating the world and giving it life. The Aztecs, Egyptians, and Hindus all had their version of the sun god.

But during the Renaissance, some scientists tried to examine the sun through the lens of physics. If the sun were made of wood or oil, then it would have exhausted its fuel long ago. And if the vast reaches of outer space did not have air, then the sun's flames would have been extinguished long ago. So the sun's eternal energy was a mystery.

In 1842, a grand challenge was issued to the scientists of the world. The French philosopher Auguste Comte, the founder of the philosophy of positivism, declared that science was indeed powerful, revealing many of the secrets of the universe, but there would be one thing that would forever be beyond the reach of science. Even the greatest scientists would never answer the question: What are the planets and the sun made of?

This was a reasonable challenge, since the bedrock of science is testability. All discoveries of science have to be reproducible and tested in the lab, but it was clearly impossible to capture sun material in a bottle and bring it back to Earth. Hence, this answer would forever be beyond our grasp.

Ironically, a few years after he made this claim in his book *The Positive Philosophy,* physicists met the challenge. The sun was primarily hydrogen.

Comte had made a slight but crucial mistake. Yes, science must always be testable, but, as we've established, most science is actually done indirectly.

Joseph von Fraunhofer was a nineteenth-century scientist who answered Comte by designing the most precise and accurate spectrographs of his time. (In a spectrograph, substances are heated up until they begin to glow with blackbody radiation. The light is then sent through a prism, where it creates a rainbow. Inside the band of colors, there are dark bands. These bands are created because electrons make quantum jumps from orbit to orbit, releasing and absorbing specific amounts of energy. Since each element creates its own characteristic bands, then each spectral band is like a fingerprint, allowing you to determine what this substance is made of. Spectrographs have also solved numerous crimes, by being able to identify where the mud of a criminal's footprint came from or the nature of the toxins found in poison or the origin of microscopic fibers and hairs. Spectrographs allow you to re-create a crime scene by determining the chemical composition of everything present.)

By analyzing the bands of light from the sun, Fraunhofer and others could tell that the sun was mainly made of hydrogen. (Strangely, physicists also found a new unknown substance in the sun. They named it helium, meaning "metal from the sun." So helium was actually first found in the sun, rather than on Earth. Later, scientists realized that helium was a gas, not a metal.)

But Fraunhofer made another important discovery. By analyzing the light from stars, he found that they were made of the same substances commonly found on the Earth. This was a profound discovery, since it indicated that the laws of physics were the same not just in the solar system but throughout the entire universe.

Once Einstein's theories had gained traction, physicists like Hans Bethe put it all together to determine what fuels the sun. If the sun is made of hydrogen, its immense gravity field can compress hydrogen until the protons fuse, creating helium and the higher elements. Since helium weighs a bit less than the protons and neutrons that combine to form it, this means that the missing mass went into energy, via Einstein's formula $E = mc^2$.

QUANTUM MECHANICS AND WAR

While physicists were debating the mind-bending paradoxes of the quantum theory, war clouds were gathering on the horizon. Adolf Hitler seized power in Germany in

1933, and waves of physicists were forced to flee Germany, be arrested, or worse.

One day, Schrödinger was witnessing Nazi brownshirts harassing innocent Jewish bystanders and shopkeepers. When he tried to stop them, they turned on him and began to beat him. They finally stopped when one of the brownshirts recognized that the person they were beating had won the Nobel Prize in Physics. Shaken, Schrödinger would soon leave Austria. Alarmed by the daily reports of repression, the best and brightest of German science left their country.

Planck, the father of the quantum theory, was ever the diplomat, and even personally pleaded with Hitler to stop the mass exodus of German scientists, which was bleeding the country of its finest minds. But Hitler simply yelled and screamed at Planck, denouncing the Jews. Afterward, Planck wrote that "it was impossible to talk to such a man." (Sadly, Planck's own son tried to assassinate Hitler, for which he was brutally tortured and then executed.)

For decades, Einstein was asked whether his equation could unleash fabulous amounts of energy locked inside the atom. Einstein would always say no, that the energy released by one atom is too small to be of practical use.

Hitler, however, wanted to use German superiority in science to create powerful weapons that the world had never seen before, weapons of terror, like the V-1 and V-2 rockets and the atomic bomb. After all, if the sun was powered by nuclear energy, then it might be possible to create a superweapon using the same source of power.

The key insight into how to exploit Einstein's equation came from physicist Leo Szilard. German physicists had shown that the uranium atom, when hit by neutrons, could split in half, releasing more neutrons. The energy released by the splitting of a single uranium atom was extremely tiny, but Szilard realized you could magnify the power of the uranium atom via a chain reaction: splitting one uranium atom released two neutrons. These neutrons could then fission two more uranium atoms, releasing four neutrons. Then you would have eight, sixteen, thirty-two, sixty-four (and so on) neutrons—that is, an exponential rise in the number of split uranium atoms, eventually creating enough energy to level a city.

Suddenly, the arcane discussions that divided the physicists at the Solvay Conference became a question of life-and-death urgency, with the fate of entire populations, nations, and civilization itself at stake.

Einstein was horrified when he learned that in Bohemia the Nazis were sealing off the pitchblende mines that contained uranium. Although a pacifist, Einstein felt compelled to write a fateful letter to President Franklin Roosevelt, urging the United States to build an atomic bomb. Roosevelt subsequently authorized the largest scientific project in history, the Manhattan Project.

Back in Germany, Werner Heisenberg, arguably the most

prominent quantum physicist on the planet, was appointed to be the head of the Nazi atomic bomb project. According to some historians, so great was the fear that Heisenberg might beat the Allies to an atomic bomb that the OSS, the forerunner of the CIA, hatched a plan to assassinate him. In 1944, a former Brooklyn Dodgers catcher, Moe Berg, was given the job. Berg attended a talk Heisenberg gave in Zurich, with orders to kill the physicist if Berg thought that the German bomb effort was near completion. (This story is elaborated in Nicholas Dawidoff's book *The Catcher Was a Spy*.)

Fortunately, the Nazi bomb project was considerably behind the Allied effort. It was underfunded, chronically late, and its base was also being bombed by the Allies. Most important, Heisenberg had not yet solved a crucial problem in making the atomic bomb: determining the amount of enriched uranium or plutonium necessary to create a chain reaction, an amount known as the critical mass. (The actual amount is roughly twenty pounds of uranium-235, which could be held in the palm of your hand.)

After the war, the world began to slowly learn that the arcane, obscure equations of the quantum theory held not only the key to atomic physics, but also perhaps to the destiny of the human race itself.

Physicists, however, began to slowly return to the question that had puzzled them before the war: how to create a complete quantum theory of matter.

4

THEORY OF ALMOST EVERYTHING

fter the war, Einstein, the towering figure who had unlocked the cosmic relationship between matter and energy and discovered the secret of the stars, found himself lonely and isolated.

Almost all recent progress in physics had been made in the quantum theory, not in the unified field theory. In fact, Einstein lamented that he was viewed as a relic by other physicists. His goal of finding a unified field theory was considered too difficult by most physicists, especially when the nuclear force remained a total mystery.

Einstein commented, "I am generally regarded as a sort of petrified object, rendered blind and deaf by the years. I find this role not too distasteful, as it corresponds fairly well with my temperament."

In the past, there was a fundamental principle that guided Einstein's work. In special relativity, his theory had to remain the same when interchanging *X, Y, Z,* and *T.* In general relativity, it was the equivalence principle, that gravity and acceleration could be equivalent. But in his quest for the theory of everything, Einstein failed to find a guiding principle. Even today, when I go through Einstein's notebooks and calculations, I find plenty of ideas but no guiding principle. He himself realized that this would doom his ultimate quest. He once observed sadly, "I believe that in order to make real progress, one must again ferret out some general principle from nature."

He never found it. Einstein once bravely said that "God is subtle, but not malicious." In his later years, he became frustrated and concluded, "I have second thoughts. Maybe God *is* malicious."

Although the quest for a unified field theory was ignored by most physicists, every now and then, someone would try their hand at creating one.

Even Erwin Schrödinger tried. He modestly wrote to Einstein, "You are on a lion hunt, while I am speaking of rabbits." Nevertheless, in 1947 Schrödinger held a press conference to announce his version of the unified field theory. Even Ireland's prime minister, Éamon de Valera, showed up. Schrödinger said, "I believe I am right. I shall look an awful fool if I am wrong." Einstein would later tell Schrödinger that he had also considered this theory and found it to be incorrect. In addition, his theory could not explain the nature of electrons and the atom.

Werner Heisenberg and Wolfgang Pauli caught the bug too, and proposed their version of a unified field theory. Pauli was the biggest cynic in physics and a critic of Einstein's program. He was famous for saying, "What God has torn asunder, let no man put together"—that is, if God had torn apart the forces in the universe, then who were we to try to put them back together?

In 1958, Pauli gave a talk at Columbia University explaining the Heisenberg-Pauli unified field theory. Bohr was in the audience. After his talk, Bohr stood up and said, "We in the back are convinced your theory is crazy. What divides us is whether your theory is crazy enough."

This began a heated discussion, with Pauli claiming that his theory was crazy enough to be true, while others said his theory was not crazy enough. Physicist Jeremy Bernstein was in the audience, and he recalled, "It was an uncanny encounter of two giants of modern physics. I kept wondering what in the world a non-physicist visitor would have made of it."

Bohr was right; the theory presented by Pauli would later be shown to be incorrect.

But Bohr had actually hit upon something important. All the easy, obvious theories had already been tried by Einstein and his associates, and they all failed. Therefore, the true unified field theory must be radically different from all previous approaches. It must be "crazy enough" to qualify as a true theory of everything.

QED

The real progress in the postwar era was made in developing a complete quantum theory of light and electrons, called quantum electrodynamics, or QED. The goal was to combine Dirac's theory of the electron with Maxwell's theory of light, thereby creating a theory of light and electrons that obeyed quantum mechanics and special relativity. (A theory combining Dirac electrons with general relativity, however, was considered to be much too difficult.)

Back in 1930, Robert Oppenheimer (who would later lead the project to build the atomic bomb) realized something profoundly disturbing. When one tried to describe the quantum theory of an electron interacting with a photon, one found that the quantum corrections actually diverged, yielding useless, infinite results. Quantum corrections were supposed to be small—that had been the guiding principle for decades. So there was an essential flaw in simply combining the Dirac equation of electrons and Maxwell's theory of photons. This haunted physicists for nearly two decades. Many physicists worked on this problem, but little progress was made.

Finally, in 1949, three young physicists working independently, Richard Feynman and Julian Schwinger in the United States, and Shin'Ichiro Tomonaga in Japan, cracked this long-standing problem.

They were spectacularly successful, able to compute

things like the magnetic property of the electron with enormous accuracy. But the way they did it was controversial and still causes physicists some unease and consternation even today.

They started with the Dirac equation and Maxwell's equation, where the mass and charge of the electron are given certain initial values (called the "bare mass and bare charge"). Then they calculated the quantum corrections to the bare mass and charge. These quantum corrections were infinite. This was the problem found earlier by Oppenheimer.

But the magic occurs here. If we assume that the original bare mass and charge were actually infinite to start with, and then calculate the infinite quantum corrections, we find that these two infinite numbers can cancel each other out, leaving a finite result! In other words, *infinity minus infinity equals zero!*

This was a crazy idea, but it worked. The strength of the magnetic field of the electron could be calculated using QED to an astonishing accuracy—that is, one part in one hundred billion.

"The numerical agreement between theory and experiment here is perhaps the most impressive in all science," Steven Weinberg noted. It is like calculating the distance from Los Angeles to New York to within the diameter of a hair. Schwinger was so proud of this that he had the symbol for this result carved on his gravestone.

This method is called renormalization theory. The procedure, however, is arduous, complex, and mind-numbing. Literally thousands of terms have to be computed exactly, and they all have to cancel precisely. The tiniest error in this thick book of equations can throw off the entire calculation. (It is no exaggeration to say that some physicists spend their entire lives calculating quantum corrections to the next decimal place using renormalization theory.)

Because the process of renormalization is so difficult, even Dirac, who helped to create QED in the first place, did not like it. Dirac felt that it seemed totally artificial, like brushing things under the rug. He once said, "This is just not sensible mathematics. Sensible mathematics involves neglecting a quantity when it turns out to be small—not neglecting it just because it is infinitely great and you do not want it!"

Renormalization theory, which could combine Einstein's special relativity with Maxwell's electromagnetism, is indeed supremely ugly. One has to master an encyclopedia of mathematical tricks in order to cancel thousands of terms. But you cannot argue with results.

APPLICATIONS OF THE QUANTUM REVOLUTION

This, in turn, paved the way for a remarkable set of discoveries, which would bring about the third great revolution in history, the high-tech revolution, including transistors and lasers, and thus help create the modern world.

Consider the transistor, perhaps the pivotal invention of the last hundred years. The transistor ushered in the information revolution, with a vast network of telecommunications systems, computers, and the internet. A transistor is basically a gate that controls the flow of electrons. Think of a valve. With a slight turn of a valve, we can control the flow of water in a pipe. In the same way, a transistor is like a tiny electronic valve that allows a small amount of electricity to control the much larger flow of electrons in a wire. Thus, a small signal can be amplified.

Similarly, the laser, one of the most versatile optical devices in history, is another by-product of the quantum theory. To create a gas laser, start with a tube of hydrogen and helium. Then inject energy into it (by applying an electric current). This sudden injection of energy causes trillions of electrons in the gas to jump to a higher energy level. However, this array of energized atoms is unstable. If one electron decays to a lower level, it releases a photon of light, which hits a neighboring pumped-up atom. This causes the second atom to decay and release another photon. Quantum mechanics predicts that the second photon vibrates in unison with the first. Mirrors can be placed at either end of the tube, magnifying this flood of photons. Eventually, this process causes a gigantic avalanche of photons, all vibrating back and forth between the mirrors in unison, creating the laser beam.

Today, lasers can be found everywhere: grocery checkout counters, hospitals, computers, rock concerts, satellites in space, etc. Not only can vast amounts of information be carried on laser beams, you can also transmit colossal amounts of energy, sufficient to burn through most materials. (Apparently, the only limitation to the energy of a laser is the stability of the lasing material and the energy that drives the laser. So, with the appropriate lasing substance and power source, one could in principle create a laser beam similar to the ones seen in science fiction movies.)

WHAT IS LIFE?

Erwin Schrödinger was a pivotal figure in formulating quantum mechanics. But Schrödinger was also interested in another scientific problem that had fascinated and dogged scientists for centuries: What is life? Could quantum mechanics answer this age-old mystery? He believed that one by-product of the quantum revolution would be the key to understanding the origin of life.

Throughout history, scientists and philosophers believed that there was some sort of life force that animated living things. When a mysterious soul entered a body, it suddenly became animate and acted human. Many believed in something called dualism, where the material body coexisted with a spiritual soul.

Schrödinger, however, believed that the code of life was hidden inside some master molecule that obeyed the laws of quantum mechanics. Einstein, for example, banished the ether from physics. Likewise, Schrödinger would try to banish the life force from biology. In 1944, he wrote a pioneering book, *What Is Life?*, that had a profound effect on a new generation of postwar scientists. Schrödinger proposed to use quantum mechanics to answer the most ancient of questions about life. In that book, he saw that a genetic code was somehow being transported from one generation of living organisms to the next. He believed that this code was stored not in a soul but in the arrangement of molecules in our cells. Using quantum mechanics, he theorized about what this mysterious master molecule could be. Unfortunately, not enough was known about molecular biology in the 1940s to answer this question.

But two scientists, James D. Watson and Francis Crick, read the book and were fascinated by the search for this master molecule. Watson and Crick realized that molecules were so small that it was impossible to see or manipulate one. This is because the wavelength of visible light is much larger than a molecule. But they had another quantum trick up their sleeve: X-ray crystallography. The wavelength of X-rays is comparable in size to molecules, so by shining

X-rays at a crystal of organic materials, the X-rays would be scattered in many directions. But the pattern of the scatter contained information about the detailed atomic structure of the crystal. Different molecules produced different X-ray patterns. A skilled quantum physicist, by looking at photographs of the scatter, could then surmise what the structure of the original molecule was. So although you could not see the molecule itself, you could decipher its structure.

Quantum mechanics was so powerful that one could determine the angle at which different atoms bound together to create molecules. Like a child playing with Tinkertoys or Legos, one could then build up, atom for atom, chains of these atoms stuck together to reproduce the actual structure of a complex molecule. Watson and Crick realized that the DNA molecule was one of the main constituents of the nucleus of a cell, so that was a likely target. By analyzing the crucial X-ray photos taken by Rosalind Franklin, they were able to conclude that the structure of the DNA molecule was a double helix.

In one of the most important papers published in the twentieth century, Watson and Crick were able to use quantum mechanics to decode the entire structure of the DNA molecule. It was a masterpiece. They demonstrated conclusively that the fundamental process of living things—reproduction—could be duplicated at the molecular level. Life was encoded on the strands of DNA found in every cell.

That was the breakthrough that made it possible to achieve the holy grail of biology, the Human Genome Project, which has given us a complete atomic description of a person's DNA.

As Charles Darwin had predicted in the previous century, it was now possible to construct the family tree of life on Earth, with every living thing or fossil a member of one branch of this tree. All of this was the product of quantum mechanics.

So the unification of the laws of quantum physics not only revealed the secrets of the universe, it also unified the tree of life.

THE NUCLEAR FORCE

We recall that Einstein was unable to complete his unified field theory, in part because he was missing a huge piece of the puzzle, the nuclear force. Back in the 1920s and 1930s, almost nothing was known about it.

But in the postwar era, buoyed by the astounding success of QED, physicists turned their attention to the next burning problem—applying the quantum theory to the nuclear force. This would be a difficult and arduous task, since they were starting from scratch and needed entirely new powerful instruments to find their way in this unknown territory.

There are two kinds of nuclear forces, the strong and the weak. Since the proton has positive charge, and since positive charges repel each other, the nucleus of the atom might ordinarily fly apart. What holds the nucleus together, overcoming electrostatic repulsion, are the nuclear forces. Without them, our entire world would dissolve into a cloud of subatomic particles.

The strong nuclear force is sufficient to keep the nucleus of many chemical elements stable indefinitely. Many have been stable since the beginning of the universe itself, especially if the number of protons and neutrons are in balance. However, some nuclei are unstable for a number of reasons, especially if they have too many protons or neutrons. If they have too many protons, then the electric repulsion will cause the nucleus to fly apart. If the nucleus has too many neutrons, then their instability can cause it to decay. In particular, the weak nuclear force is not strong enough to hold the neutron together permanently, so eventually it falls apart. For example, half of any collection of free neutrons will decay in fourteen minutes. What is left are three particles: the proton; the electron; and another mysterious new particle, the anti-neutrino, which we will discuss later.

Studying the nuclear force is exceedingly difficult, since the nucleus is about one hundred thousand times smaller than an atom. To probe inside a proton, physicists needed a new tool, the particle accelerator. We saw how years before Ernest Rutherford used the rays emitted by radium encased in a lead box to discover the nucleus. To explore deeper inside the nucleus, physicists needed even more powerful sources of radiation.

In 1929, Ernest Lawrence invented the cyclotron, the forerunner of the giant particle accelerators of today. The basic principle behind the cyclotron is simple. A magnetic field forces protons to move in a circular path. At each cycle, the protons are given a small boost of energy by an electric field. Eventually, after many revolutions, the beam of protons can reach millions and even billions of electron volts. (The basic principles of a particle accelerator are so straightforward that I even built an electron particle accelerator, a betatron, when I was in high school.)

This beam, in turn, is eventually directed at a target, where it smashes into other protons. By sifting through the enormous debris from this collision, scientists were able to identify new, previously undiscovered particles. (This process of shooting beams of particles to smash protons apart is a clumsy, imprecise operation. It has been compared to throwing a piano out the window, and then trying to determine all the piano's properties by analyzing the sound of the crash. As clumsy as this process is, it is one of the only ways we have to probe the interior of the proton.)

When physicists first smashed protons with a particle accelerator in the 1950s, they found, to their dismay, an entire zoo of unexpected particles.

It was an embarrassment of riches. Nature, it was believed, was supposed to become simpler the deeper you searched, not more complex. To the quantum physicist, it seemed that perhaps nature really was malicious after all.

Frustrated by this flood of new particles, Robert Oppenheimer declared that the Nobel Prize in Physics should be given to the physicist who did *not* discover a new particle that year. Enrico Fermi declared that "if I had known there would be so many particles with Greek names, I would have become a botanist rather than a physicist."

Researchers were drowning in subatomic particles. The mess prompted some physicists to claim that perhaps the human mind was not smart enough to understand the subatomic realm. After all, they argued, it is impossible to teach a dog calculus, so perhaps the human mind is not powerful enough to understand what's happening in the nucleus of an atom.

Some of the confusion began to be clarified with the work of Murray Gell-Mann and his colleagues at the California Institute of Technology (Caltech), who claimed that, inside the proton and neutron, there were three even smaller particles called quarks.

It was a simple model, but it worked spectacularly well in arranging the particles into groups. Like Mendeleyev before him, Gell-Mann could predict the properties of new strongly interacting particles by looking at the gaps in his theory. In 1964, another particle predicted by the quark model, called the omega-minus, was actually found, verifying the basic correctness of this theory, for which Gell-Mann won the Nobel Prize.

The reason the quark model was able to unify so many particles is because it was based on a symmetry. Einstein, we recall, introduced a four-dimensional symmetry that turned space into time and vice versa. Gell-Mann introduced equations containing three quarks; when you interchanged them inside an equation, the equation remained the same. This new symmetry described the reshuffling of three quarks.

POLAR OPPOSITES II

The other great physicist at Caltech, Richard Feynman, who renormalized QED, and Murray Gell-Mann, who introduced the quark, were polar opposites in their personality and temperament.

In the popular media, physicists are universally portrayed either as mad scientists (like Doc Brown in *Back to the Future*) or hopelessly inept nerds, as in *The Big Bang Theory*. However, in actuality physicists come in all shapes and sizes and personality types.

Feynman was a colorful gadfly, ever the showman and the clown, full of ribald stories of his outrageous stunts, told in a rough working-class accent. (During World War II, he once cracked the safe containing the secrets of the atomic bomb at the Los Alamos National Laboratory. Inside the safe, he left a cryptic note. When officials found this note the next day, it set off a major alarm and panic at the nation's top secret laboratory.) Nothing was too unorthodox or outrageous for Feynman; out of curiosity, he once even sealed himself in a hyperbaric chamber to see if he could have an out-of-body experience.

Gell-Mann, however, was the opposite, ever the gentleman, precise in his words and manners. Bird-watching, collecting antiques, linguistics, and archaeology were his favorite pastimes, not reciting hilarious stories. But as different as they were in character, they both had the same drive and determination, which helped them to penetrate the mysteries of the quantum theory.

WEAK FORCE AND GHOSTLIKE PARTICLES

Meanwhile, great strides were being made in understanding the weak nuclear force as well, which is about a million times weaker than the strong force.

The weak force, for example, is not powerful enough to hold the nuclei of many types of atoms together, so they fall apart and decay into smaller subatomic particles. Radioactive decay, as we have seen, is the reason the inside of the Earth is so hot. The fierce energy of thundering volcanoes and terrible earthquakes comes from the weak nuclear force. A new particle had to be introduced to explain the weak force. A neutron, for example, is unstable and eventually decays into a proton and an electron. This is called beta decay. But in order for the calculations to work out, physicists needed to introduce a third particle, a shadowy particle called the neutrino.

The neutrino is sometimes called the ghost particle, because it can penetrate entire planets and stars without being absorbed. At this very instant, your body is being radiated by a flood of neutrinos from deep space, some of which traveled through the entire planet Earth. In fact, some of these neutrinos could penetrate a block of solid lead that stretches from the Earth to the nearest star.

Pauli, who predicted the existence of the neutrino in 1930, once lamented, "I have committed the ultimate sin. I have introduced a particle that can never be observed." As elusive as this particle is, it was finally discovered experimentally in 1956 by analyzing the intense radiation emitted from a nuclear power plant. (Although the neutrino hardly interacts with ordinary matter, physicists compensated for this by exploiting the vast number of neutrinos emitted in a nuclear reactor.)

To make sense of the weak nuclear force, physicists once again introduced a new symmetry. Since the electron and neutrino were a pair of weakly interacting particles, it was proposed that they could be paired, giving us a symmetry. Then this new symmetry, in turn, could be coupled to the older symmetry of Maxwell's theory. The resulting theory was called the electroweak theory, which unified electromagnetism with the weak nuclear force.

This electroweak theory of Steven Weinberg, Sheldon Glashow, and Abdus Salam won them the Nobel Prize in 1979.

So light, instead of being united with gravity, as Einstein had hoped, actually preferred to be united with the weak nuclear force.

Thus, the strong force was based on Gell-Mann's symmetry, which binds the three quarks together to make protons and neutrons, while the weak nuclear force was based on a smaller symmetry, the rearranging of the electron with the neutrino, which is then combined with electromagnetism.

But powerful as the quark model and the electroweak theory were in describing the zoo of subatomic particles, this still left a huge gap. The burning question was: What holds all these particles together?

YANG-MILLS THEORY

Because the Maxwell field had so much success in predicting the properties found in electromagnetism, physicists began to study a new, more powerful version of Maxwell's equation. It was proposed by Chen Ning Yang and Robert L. Mills in 1954. Instead of just one field, written down by Maxwell in 1861, it introduced a family of these fields. The same symmetry that Gell-Mann used to rearrange the quarks in this theory was now used to rearrange this new collection of Yang-Mills fields into one another.

The idea was simple. What holds the atom together is the electric field, which is described by Maxwell's equations. Then perhaps what holds the quarks together is a generalization of Maxwell's equations—that is, the Yang-Mills fields. The same symmetry that describes the quarks is now applied to the Yang-Mills field.

However, for several decades, this simple idea languished because, when calculating the properties of the Yang-Mills particles, the result was again infinite, just like we saw in QED. Unfortunately, the bag of tricks introduced by Feynman was not enough to renormalize the Yang-Mills theory. For years, physicists despaired of finding a finite theory of the nuclear force.

Finally, an enterprising Dutch grad student, Gerard 't Hooft, had the courage and raw stamina to plow through this thicket of infinite terms and, via brute force, renormalize the Yang-Mills field. By then, computers were advanced enough to analyze these infinities. When his computer program spit out a series of zeros representing these quantum corrections, he knew he must be right.

News of this breakthrough caught the immediate attention of physicists. Physicist Sheldon Glashow would exclaim, "Either this guy's a total idiot, or he's the biggest genius to hit physics in years!"

It was a tour de force that would win 't Hooft and his adviser, Martinus Veltman, the Nobel Prize in 1999. Suddenly, there was a new field that could be used to bind together the known particles in the nuclear force and explain the weak force. When applied to quarks, the Yang-Mills field was called the gluon, because it acted like a glue to bind the quarks together. (Computer simulations show that the Yang-Mills field condenses into a taffy-like substance, which then holds the quarks together, like glue.) To do this, one needed quarks coming in three types, or colors, obeying Gell-Mann's three-quark symmetry. So a new theory of the strong force began to gain wide acceptance. This new theory was christened quantum chromodynamics (QCD), and today this represents the best-known representation of the strong nuclear force.

HIGGS BOSON—THE GOD PARTICLE

So gradually, a new theory was emerging from all this chaos, called the Standard Model. The confusion surrounding the zoo of subatomic particles was lifting. The Yang-Mills field (called the gluon) held the quarks together in the neutron and proton, and another Yang-Mills field (called the W and Z particles) described the interaction between electrons and neutrinos.

But what prevented final acceptance of the Standard Model was the lack of the final piece of the jigsaw puzzle of particles, called the Higgs boson, or sometimes the God particle. Symmetry was not enough. We need a way to break that symmetry because the universe we see around us is not perfectly symmetrical.

When we look at the universe today, we see the four forces all working independently of one another. Gravity, light, and the nuclear forces, at first glance, seem to have nothing in common. But as you go back in time, these forces begin to converge, perhaps leaving only one force at the instant of creation.

A new picture began to develop that used particle physics to explain the greatest mystery of cosmology, the birth of the universe. Suddenly, two very different fields, quantum mechanics and general relativity, began to gradually turn into one.

In this new picture, at the instant of the Big Bang, all the four forces were merged into a single superforce that obeyed the master symmetry. This master symmetry could rotate all the particles of the universe into one another. The equation that governed the superforce was the God equation. Its symmetry was the symmetry that had eluded Einstein and physicists ever since.

After the Big Bang, as the universe expanded, it began to cool and the various forces and symmetries began to break into pieces, leaving the fragmented weak and strong force symmetries of the Standard Model of today. This process is called symmetry breaking. This means that we need a mechanism that can precisely break this original symmetry, leaving us with the Standard Model. That is where the Higgs boson comes in.

To imagine this, think of a dam. The water in the reservoir also has a symmetry. If you rotate the water, the water looks very much the same. We all know from experience that water runs downhill. This is because, according to Newton, water always seeks out a lower energy state. If the dam were to break, the water would suddenly rush downstream into a lower energy state. So the water behind the dam is in a higher energy state. Physicists call the state of the water behind the dam the false vacuum, because it is unstable until the water in the burst dam reaches the true vacuum, meaning the lowest energy state in the valley below. After the dam bursts, the original symmetry is gone, but the water has reached its true ground state.

This effect is also found when you analyze water that is beginning to boil. Just before it boils, the water is in the false vacuum. It is unstable but symmetrical—that is, you can rotate the water and the water looks the same. But eventually, tiny bubbles form, where each bubble exists in a lower energy state than the surrounding water. Each bubble starts to expand, until enough bubbles merge and the water boils.

According to this scenario, the universe was originally in a perfectly symmetrical state. All the subatomic particles were part of the same symmetry, and they all had zero mass. Because they had zero mass, they could be rearranged but the equation would remain the same. However, for some unknown reason, it was unstable; it was in the false vacuum. The field necessary to shift to the true (but broken) vacuum is the Higgs field. Like Faraday's electric field that permeated all corners of space, the Higgs field also filled up all of space-time.

But for some reason, the symmetry of the Higgs field began to break.

Tiny bubbles began to form inside the Higgs field. Outside the bubbles, all particles remained massless and symmetrical. Inside the bubble, some particles had mass. As the Big Bang progressed, the bubble expanded rapidly, the particles began to acquire different masses, and the original symmetry was broken. Eventually, the entire universe exists in the new vacuum state inside a gigantic bubble.

So by the 1970s, the hard work of scores of physicists began to pay off. After decades of wandering in the wilderness, they were finally beginning to fit all the pieces of the jigsaw puzzle together. They realized that by cobbling together three theories (representing the strong, weak, and electromagnetic forces) they could write a set of equations that truly coincided with the results observed in the laboratory.

The key was to create a master symmetry by gluing together three distinct smaller symmetries. The first symmetry described the strong nuclear force, which shuffled three quarks among each other. The second symmetry described the weak force, by shuffling electrons and neutrinos. The third symmetry described the original Maxwell field. The final theory was awkward, but it was hard to argue with success.

THEORY OF ALMOST EVERYTHING

Remarkably, the Standard Model could accurately predict the properties of matter all the way back to a fraction of a second after the Big Bang.

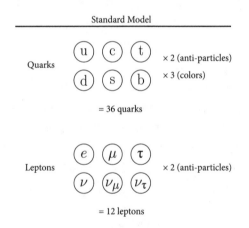

Figure 9. The Standard Model is a strange collection of subatomic particles that accurately describes our quantum universe, with thirty-six quarks and anti-quarks, twelve weakly interacting particles and anti-particles (called leptons), and a large assortment of Yang-Mills fields and Higgs bosons, particles that are created when you excite the Higgs field.

Although the Standard Model represented our best understanding of the subatomic world, there were numerous glaring holes. First, the Standard Model made no mention of gravity. This was a huge problem, since gravity is the force that controls the large-scale behavior of the universe. And every time physicists tried to add it to the Standard Model, they could not solve the equations. The quantum corrections due to it, instead of being small, turn out to be infinite, just like QED and Yang-Mills particles. So the Standard Model is unable to shed light on some of the stubborn secrets of the universe, such as what happened before the Big Bang and what lies inside a black hole. (We will return to these important questions later.)

Second, the Standard Model was created by splicing together by hand the theories that described the various forces, so the resulting theory was a patchwork. (One physicist compared it to taping a platypus, an aardvark, and a whale together and declaring it to be nature's most elegant creature. The resulting animal, it was said, was one only a mother could love.)

Third, the Standard Model had a number of parameters that were undetermined (such as the masses of the quarks and the strength of the interactions). In fact, there are about twenty constants that had to be put in by hand, with no understanding whatsoever of where these constants came from or what they represented.

Fourth, it had not just one copy but three identical copies, or generations, of the quarks, gluons, electrons, and neutrinos in the Standard Model. (So altogether, there are thirty-six quarks, with three colors, three generations, along with their corresponding anti-particles, and twenty free parameters.) Physicists found it difficult to believe that anything so clumsy and unwieldy could be the fundamental theory of the universe.

LHC

Because there is so much at stake, nations are willing to spend billions to create the next-generation particle accelerators. Currently, the headlines have been dominated by the Large Hadron Collider outside Geneva, Switzerland, the largest machine of science ever built, costing more than $12 billion and stretching almost seventeen miles in circumference.

LHC looks like a huge doughnut that straddles the border between Switzerland and France. Inside the tube, protons are accelerated until they reach extremely high energy. Then they collide with another high-energy beam of protons heading in the opposite direction, releasing fourteen trillion electron volts of energy and creating an enormous shower of subatomic particles. The world's most advanced computers are then used to make sense out of this cloud of particles.

The goal of the LHC is to duplicate the conditions found shortly after the Big Bang and thereby to create these unstable particles. Finally, in 2012, the Higgs boson, the last piece of the Standard Model, was found.

Although this was a great day for high-energy physics, physicists realized there was still a long way to go. On one hand, the Standard Model does describe all particle interactions, from deep inside the proton to the very edges of the visible universe. The problem is that the theory is ungainly. In the past, every time physicists probed the fundamental nature of matter, new and elegant symmetries began to emerge, so physicists found it problematic that, at the most fundamental level, nature seemed to prefer a slapdash theory.

In spite of its practical successes, it is obvious to everyone that the Standard Model is just a warm-up act for the final theory, yet to come.

Meanwhile, physicists, buoyed by the astonishing successes of the quantum theory when applied to subatomic particles, began to reexamine the theory of general relativity, which had languished for decades. Now physicists set their eyes on a more ambitious goal—to combine the Standard Model with gravity, meaning that one would need a quantum theory of gravity itself. This would truly be a theory of everything, where all quantum corrections to both the Standard Model and general relativity could be calculated.

Previously, renormalization theory was a clever sleight of

hand that canceled all the quantum corrections of QED and the Standard Model. The key was to represent the electromagnetic and the nuclear forces as particles, called photons and Yang-Mills particles, and then magically wave your hand to make the infinities disappear by reabsorbing them elsewhere. All the unpleasant infinities were brushed under the rug.

Naively, physicists followed this time-honored tradition and took Einstein's theory of gravity and introduced a new point particle of gravity, called the graviton. So the smooth surface introduced by Einstein to represent the fabric of space-time was now surrounded by a cloud of trillions of tiny graviton particles.

Sadly, the bag of tricks painfully accumulated by physicists for the past seventy years to eliminate these infinities failed for the graviton. The quantum corrections created by gravitons were infinite and could not be reabsorbed somewhere else. Here, physicists hit a brick wall. Their winning streak came to an abrupt end.

Frustrated, physicists then began to try a more modest goal. Unable to create a complete quantum theory of gravity, they tried to calculate what happens when ordinary matter is quantized, leaving gravity alone. This meant calculating the quantum corrections due to stars and galaxies but keeping gravity untouched. By only quantizing the atom, it was hoped to create a stepping-stone and gain insight into the larger goal of formulating a quantum theory of gravity.

This was a more modest goal, but it opened the floodgates to an astonishing array of new, fascinating physical phenomena that would challenge the way we view the universe. Suddenly, quantum physicists encountered the most bizarre phenomena in the universe: black holes, wormholes, dark matter and dark energy, time travel, and even the creation of the universe itself.

But the discovery of these strange cosmic phenomena was also a challenge for the theory of everything that must now explain not only the familiar subatomic particles of the Standard Model but all these strange phenomena that stretch the human imagination.

5

THE DARK UNIVERSE

I n 2019, newspapers and websites across the planet splashed sensational news on the front page: astronomers had just taken the first photograph of a black hole. Billions of people saw the stark image, a red ball of hot fiery gas with a black, round silhouette in the middle. This mysterious object captured the public's imagination and dominated the news. Not only have black holes intrigued and fascinated physicists, but they have also entered into the public's consciousness, being featured in numerous science specials and a plethora of movies.

The black hole that was photographed by the Event Horizon Telescope lies inside the galaxy M87, 53 million light-years from Earth. The black hole is truly a monster, weighing in at a staggering five billion times the mass of the sun. Our entire solar system, even past Pluto, could easily fit inside the black silhouette in the photograph.

To accomplish this stunning achievement, astronomers created a super telescope. Normally, a radio telescope is not large enough to take in enough faint radio signals to create an image of an object so distant. But astronomers were able to photograph this black hole by lashing together the signals from five individual ones scattered around the world. By using supercomputers to carefully combine these diverse signals, they effectively created a single giant radio telescope the size of planet Earth itself. This composite was so powerful that it could, in principle, detect an orange sitting on the surface of the moon from the Earth.

A host of new, remarkable astronomical discoveries like this have rejuvenated interest in Einstein's theory of gravity. Sadly, for the past fifty years, research in Einstein's general relativity was relatively stagnant. The equations were fiendishly difficult, often involving hundreds of variables; and experiments on gravity were simply too expensive, involving detectors that were miles across.

The irony is that, although Einstein had reservations about the quantum theory, the current renaissance in relativity research has been fueled by the merger of the two, by the application of the quantum theory to general relativity. As we mentioned, a complete understanding of the graviton and how to eliminate its quantum corrections is considered too difficult, but a more modest application of the quantum theory to stars (neglecting graviton corrections) has opened the heavens to a wave of startling scientific breakthroughs.

WHAT IS A BLACK HOLE?

The basic idea of a black hole actually can be traced back to Newton's discovery of the laws of gravity. His *Principia* gave us a simple picture: if you fire a cannonball with enough energy, it will completely circle the Earth, then return to its original point.

But what happens if you aim the cannonball straight up? Newton realized that the cannonball will eventually reach a maximum height and then fall back to Earth. But with enough energy, the cannonball would reach escape velocity—that is, the speed necessary to escape the Earth's gravity and soar into space, never to return.

It is a simple exercise, using Newton's laws to calculate the escape velocity of the Earth, which turns out to be 25,000 miles per hour. This is the velocity that our astronauts had to attain to reach the moon in 1969. If you do not reach escape velocity, then you will either enter orbit or fall back to Earth.

In 1783, an astronomer named John Michell asked himself a deceptively simple question: What happens if the escape velocity is the speed of light? If a light beam is emitted from a giant star so massive that its escape velocity is the speed of light, then perhaps even its light cannot escape. All light emitted from this star will eventually fall back into the star. Michell called these dark stars, celestial bodies that appeared black because light could not escape their immense gravity. Back in the 1700s, scientists knew little about the physics of stars and did not know the correct value for the speed of light, and hence this idea languished for several centuries.

In 1916, during World War I, German physicist Karl Schwarzschild was stationed on the Russian front as an artilleryman. While fighting in the middle of a bloody war, he found time to read and digest Einstein's famous 1915 paper introducing general relativity. In a brilliant stroke of mathematical insight, Schwarzschild somehow found an exact solution of Einstein's equations. Instead of solving the equations for a galaxy or the universe, which was too difficult, he started with the simplest of all possible objects, a tiny point particle. This object, in turn, would approximate the gravity field of a spherical star as seen from a distance. One could then compare Einstein's theory with experiment.

Einstein's reaction to Schwarzschild's paper was ecstatic. Einstein realized that this solution of his equations would allow him to make more precise calculations with his theory, such as the bending of starlight around the sun and the wobbling of the planet Mercury. So instead of making crude approximations to his equations, he could calculate exact results from his theory. This was a monumental breakthrough that would prove important for understanding black holes. (Schwarzschild died shortly after his remarkable discovery. Saddened, Einstein wrote a moving eulogy for him.)

But despite the enormous impact of Schwarzschild's solution, it also raised some bewildering questions. From the start, his solution had weird properties that pushed the boundaries of our understanding of space and time. Surrounding a supermassive star was an imaginary sphere (which he called the magic sphere and today is called the event horizon). Far outside this sphere, the gravity field resembled an ordinary Newtonian star's, so Schwarzschild's solution could be used to approximate its gravity. But if you were unfortunate enough to approach the star and pass through the event horizon, you would be trapped forever and would be crushed to death. The event horizon is the point of no return: anything that falls in never comes out.

But as you approached the event horizon, even more bizarre things would begin to happen. For example, you would encounter light beams that had been trapped for perhaps billions of years and are still orbiting the star. The gravity pulling on your feet would be greater than the gravity pulling on your head, so you would be stretched like spaghetti. In fact, this spaghettification becomes so severe that even the atoms of your body get pulled apart and eventually disintegrate.

To someone watching this remarkable event from a great distance, it would appear as if time inside the rocket ship on the edge of the event horizon had gradually slowed down. In fact, to an outsider, it appears as if time has stopped as the ship hits the event horizon. What is remarkable is that, to the astronauts in the ship, everything seems to be normal as they pass through the event horizon—normal, that is, until they are torn apart.

This concept was so bizarre that, for many decades, it was considered science fiction, a strange by-product of Einstein's equations that didn't exist in the real world. Astronomer Arthur Eddington once wrote that "there should be a law of Nature to prevent a star from behaving in this absurd way!"

Einstein even wrote a paper arguing that, under normal conditions, black holes could never form. In 1939, he showed that a whirling ball of gas could never be compressed by gravity to within the event horizon.

Ironically, that very same year, Robert Oppenheimer and his student Hartland Snyder showed that black holes could indeed form from natural processes that Einstein did not foresee. If you start with a giant star ten to fifty times more massive than our sun, when it uses up its nuclear fuel, it can eventually explode as a supernova. If the remnant of the explosion is a star that is compressed by gravity to its event horizon, then it can collapse into a black hole. (Our sun is not massive enough to undergo a supernova explosion, and its event horizon is about four miles across. No known natural process can squeeze our sun down to two miles, and hence our sun will not become a black hole.)

Physicists have discovered that there are at least two types of black holes. The first type is the remnant of a giant star as described above. The second type of black hole is found at the center of galaxies. These galactic black holes can be millions or even billions of times more massive than our sun. Many astronomers believe that black holes lie in the center of every galaxy.

In the last few decades, astronomers have identified hundreds of possible black holes in space. At the center of our own Milky Way lies a monster black hole whose mass is two to four million times that of our sun. It is located in the constellation Sagittarius. (Unfortunately, dust clouds obscure the area, so we cannot see it. But if the dust clouds were to part, then every night, a magnificent, blazing fireball of stars, with the black hole at its center, would light up the night sky, perhaps outshining the moon. It would truly be a spectacular sight.)

The latest excitement concerning black holes came about when the quantum theory was applied to gravity. These calculations unleashed a wellspring of unexpected phenomena that test the limits of our imagination. As it turns out, our guide through this uncharted territory was totally paralyzed.

As a graduate student at Cambridge University, Stephen Hawking was an ordinary youth, without much direction or purpose. He went through the motions of being a physicist, but his heart was not there. It was obvious that he was brilliant, but he seemed unfocused. But one day, he was diagnosed with amyotrophic lateral sclerosis (ALS) and told he would die within two years. Although his mind would be intact, his body would rapidly waste away, losing all ability to function, until he died. Depressed and shaken to the core, he realized that his life up to that point had been wasted.

He decided to dedicate the few remaining years of his life to doing something useful. To him, this meant solving one of the biggest problems in physics: the application of the quantum theory to gravity. Fortunately, his disease progressed much more slowly than his doctors predicted, so he was able to continue pathbreaking research in this new area even as he was confined to a wheelchair and lost control of his limbs and even vocal cords. I once was invited by Hawking to give a talk at a conference he was organizing. I had the pleasure of visiting his house and was surprised by the different gadgets that allowed him to continue his research. One device was a page turner. You could put a journal into this contraption, and it would automatically turn the pages. I was impressed by the degree to which he was determined not to allow his illness to detract from his life's goal.

Back then, most theoretical physicists were working on the quantum theory, but a small handful of renegades and diehards were trying to find more solutions to Einstein's equation. Hawking asked himself a different but profound question: What happens when you combine these two systems and apply quantum mechanics to a black hole?

He realized that the problem of calculating quantum corrections to gravity was much too difficult to solve. So he chose a simpler task: calculating quantum corrections just to the atoms inside a black hole, ignoring the more complex quantum corrections of the gravitons.

The more he read about black holes, the more he realized that something was wrong. He began to suspect that the traditional thinking—that nothing can escape a black hole—violated the quantum theory. In quantum mechanics, everything is uncertain. A black hole looks perfectly black because it absorbs absolutely everything. But perfect blackness violated the uncertainty principle. Even blackness had to be uncertain.

He came to the revolutionary conclusion that black holes must necessarily emit a very faint glow of quantum radiation.

Hawking then showed that the radiation emitted by a black hole was actually a form of blackbody radiation. He calculated this by realizing that the vacuum was not just the state of nothingness but was actually bubbling with quantum activity. In the quantum theory, even nothingness is in a state of constant, churning uncertainty, where electrons and anti-electrons could suddenly jump out of the vacuum, then collide and disappear back into the vacuum. So nothingness was actually frothing with quantum activity. He then realized that if the gravitational field was intense enough, then electron and anti-electron pairs could be created out of the vacuum, creating what are called virtual particles. If one member falls into the black hole, while the other particles escapes, it would create what is now called Hawking radiation. The energy to create this pair of particles comes from the energy contained in the black hole's gravity field. Because the second particle leaves the black hole forever, it means that the net matter and energy content of the black hole and its gravity field has decreased.

This is called black hole evaporation and describes the ultimate fate of all black holes: they will gently radiate Hawking radiation for trillions of years, until they exhaust all their radiation and die in a fiery explosion. So even black holes have a finite lifetime.

Trillions upon trillions of years from now, the stars of the universe will have exhausted all their nuclear fuel and become dark. Only black holes will survive in this bleak era. But even black holes must eventually evaporate, leaving nothing but a drifting sea of subatomic particles. Hawking asked himself another question: What happens if you throw a book into a black hole? Is the information in that book lost forever?

According to quantum mechanics, information is never lost. Even if you burn a book, by tediously analyzing the molecules of the burned paper, it's possible to reconstruct the entire book.

But Hawking stirred up a hornet's nest of controversy by saying that information thrown inside a black hole is indeed lost forever, and that quantum mechanics therefore breaks down in a black hole.

As previously mentioned, Einstein once said that "God does not play dice with the world"—that is, you cannot reduce everything to chance and uncertainty. Hawking added, "Sometimes God throws the die where you cannot find them," meaning that the dice may land inside a black hole, where the laws of the quantum may not hold. So the laws of uncertainty fail when you go past the event horizon.

Since then, other physicists have come to the defense of quantum mechanics, showing that advanced theories like string theory, which we will discuss in the next chapter, can preserve information even in the presence of black holes. Eventually, Hawking conceded that perhaps he was wrong. But he proposed his own novel solution. Perhaps when you throw a book into a black hole, the information is not lost forever, as he previously thought, but it comes back out, in the form of Hawking radiation. Encoded within the faint Hawking radiation is all the information necessary to re-create the original book. So perhaps Hawking was incorrect, but the correct solution lies in the radiation that he had found previously.

In conclusion, whether information is lost in a black hole is still an ongoing question, fiercely debated among physicists. But ultimately we may have to wait until we have the final quantum theory of gravity that includes graviton quantum corrections. In the meantime, Hawking turned to the next puzzling question involving combining the quantum theory and general relativity.

THROUGH THE WORMHOLE

If black holes eat up everything, then where does all that stuff go?

The short answer is, we don't know. The answer may ultimately be solved by unifying the quantum theory with general relativity.

Only when we finally find a quantum theory of gravity (and not just matter) can we answer this question: What lies on the other side of a black hole?

But if we blindly accept Einstein's theory, then we get into trouble, since his equations predict that the gravitational force at the very center of a black hole or the beginning of time is infinite, which makes no sense.

But in 1963, mathematician Roy Kerr found an entirely new solution to Einstein's equations for a rotating black hole. Previously, in Schwarzschild's work, black holes collapsed into a stationary, tiny dot, called a singularity, where gravitational fields became infinite and everything was crushed into a single point. But if you analyze Einstein's equations for a spinning black hole, Kerr found that strange things happen.

First, the black hole does not collapse into a dot. Instead, it collapses into a rapidly spinning ring. (Centrifugal forces on the spinning ring are strong enough to prevent the ring from collapsing under its own gravity.)

Second, if you fall through the ring, it's possible you may not be crushed to death at all but may pass through the ring. The gravity inside the ring is actually finite.

Third, the mathematics indicates that as you pass through the ring, you could enter a parallel universe. You literally leave our universe and enter into another sister universe. Think of two sheets of paper, stacked one on top of the other. And then stick a straw through both of them. By passing through the straw, you leave one universe and enter a parallel universe. This straw is called a wormhole.

Fourth, as you reenter the ring, you could proceed to another universe. Like taking an elevator in an apartment building, you pass from one floor to the next, from one universe to another. Each time you reenter the wormhole, you could enter an entirely new universe. So this introduced a startling new picture of a black hole. At the very center of a spinning black hole, we find something resembling the looking glass of Alice. On one side, we have the tranquil countryside of Oxford, England. But if you stuck your hand through the looking glass, you would wind up somewhere else entirely.

Fifth, if you succeed in passing through the ring, there is also the chance that you will wind up in a distant region of your same universe. So the wormhole could be like a subway system, taking an invisible shortcut through space and time. Calculations show that you might be able to go faster than the speed of light, or even go backward in time, perhaps without violating known physical laws.

These bizarre conclusions, no matter how outrageous, cannot be easily dismissed, since they are solutions to Einstein's equation, and they describe spinning black holes, which we now believe are by far the most common kind.

Wormholes were actually first introduced by Einstein himself in 1935, in a paper with Nathan Rosen. They imagined two black holes joined together, which resemble two funnels in space-time. If you fell into one funnel, you would be thrust out the end of the other funnel without being crushed to death.

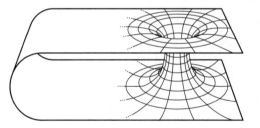

Figure 10. In principle, one might hypothetically be able to reach the stars or even the past by going through the wormhole.

There is this famous line in T. H. White's novel *The Once and Future King*: "Everything not forbidden is compulsory."

Physicists actually take this statement seriously. Unless there is a physical law against a phenomenon, perhaps it exists somewhere in the universe.

For example, even though wormholes are notoriously hard to create, some physicists have speculated that wormholes may have existed at the beginning of time and then expanded after the Big Bang. Maybe they exist naturally. One day, our telescopes may actually see a wormhole in space. Although wormholes have fired up the imagination of science fiction writers, actually creating one in a laboratory poses daunting problems.

First, you need to assemble vast amount of positive energy, comparable to a black hole, to open the gateway through space-time. This alone would require the technology of a very advanced civilization. So we don't expect amateur inventors to be able to create a wormhole in their basement laboratories any time soon.

Second, such a wormhole is going to be unstable and will close by itself, unless one adds a new, exotic ingredient, called negative matter or negative energy, which is entirely different from antimatter. Negative matter and energy are repulsive, which can keep the wormhole from collapsing.

Physicists have never seen negative matter. In fact, it would obey anti-gravity, so it would fall up, rather than down. If negative matter were on the Earth billions of years ago, it would have been repelled by the gravity of the Earth and flung into outer space. So we don't expect to find negative matter on the Earth.

Negative energy, in contrast to negative matter, does in fact exist, but only in minuscule amounts, too small to be of practical value. Only a very advanced civilization, perhaps millennia more advanced than us, would be able to harness enough positive and negative energy to create a wormhole and then keep it from collapsing.

Third, radiation from gravity itself (called graviton radiation) might be enough to cause the wormhole to explode.

Ultimately the final answer to the question of what happens when you fall into a black hole must await a true theory of everything, in which both matter and gravity are quantized.

Some physicists have seriously proposed the controversial idea that when stars fall into a black hole, they are not crushed into a singularity but instead are blown out the other side of a wormhole, creating a white hole. A white hole obeys precisely the same equations as a black hole, except the arrow of time is reversed, so matter spews out of a white hole. Physicists have looked for white holes in space, but so far have turned up empty-handed. The point of mentioning white holes is that perhaps the Big Bang was originally a white hole, and all the stars and planets we see in the heavens were flung out of a black hole—about fourteen billion years ago.

The point is that only a theory of everything can tell us what lies on the other side of a black hole. Only by calculating quantum corrections to gravity can we answer the deepest questions raised by wormholes.

But if wormholes might one day take us instantly across the galaxy, can they also take us to the past?

TIME TRAVEL

Time travel is a staple of science fiction, ever since H. G. Wells's *The Time Machine*. We can move freely in three dimensions (forward, sideways, and upward), so perhaps there was a way to move in the fourth dimension, time. Wells envisioned entering a time machine, spinning a dial, and then soaring hundreds of thousands of years into the future to the year 802,701 CE.

Since then, scientists have studied the possibility of time travel. When Einstein first proposed his theory of gravity in 1915, he was worried that his equations might allow one to twist time so that one could enter the past, which he believed would indicate a flaw in his theory. But this nagging problem became a real possibility in 1949, when his neighbor at Princeton's famed Institute for Advanced Study, the great mathematician Kurt Gödel, found that if the universe rotated, and one could travel around the spinning universe fast enough, then one could enter the past—that is, you could return before you left. Einstein was stunned by this unorthodox solution. Einstein, in his memoirs, finally concluded that even though time travel was possible in Gödel's universe, it could be dismissed "on physical grounds," meaning that the universe expanded and did not rotate.

Now, although physicists are still not convinced about the possibility of time travel, they are taking the question very seriously. A variety of solutions to Einstein's equations have been discovered that allow for time travel.

To Newton, time was like an arrow. Once fired, it would unerringly proceed with uniform speed throughout the universe. One second on the Earth was one second everywhere in space. Clocks could be synchronized anywhere in the universe. To Einstein, however, time was more like a river. It could speed up or slow down as it meandered its way across stars and galaxies. Time could tick at different rates across the universe. The new picture, however, states that the river of time could have whirlpools that might sweep you to the past (physicists call them CTCs, or closed timelike curves). Or perhaps the river of time might fork into two rivers, so the time line splits, creating two parallel universes.

Hawking was so fascinated by time travel that he issued a challenge to other physicists. He believed there must be a hidden law of physics, not yet found, that he called the chronology protection conjecture, which ruled out time travel once and for all. But try as he might, he could never prove this hypothesis. This means that time travel might still be consistent with the laws of physics, with nothing to prevent the existence of time machines.

Also, tongue in cheek, he said that time travel was not possible, because "where are the tourists from the future?" At every major historical event, there should be hordes of tourists with their cameras elbowing one another, frantically trying to get the best picture of the event to show their friends in the future.

For the moment, think of the mischief you could create if you had a time machine. Going back in time, you could make bets on the stock market and become a billionaire. You could change the course of past events. History could never be written down. Historians would be out of a job.

Time travel, of course, has serious problems. There are a host of logical paradoxes associated with time travel, such as:

- Making the present impossible: If you go back in time to meet your grandfather as a child and kill him, then how can you even exist?
- Time machine from nowhere: Someone from the future gives you the secret of time travel. Years later, you go back in time and give the secret of time travel to your younger self. Then where did the secret of time travel come from?

· Becoming your own mother: Science fiction writer Robert Heinlein wrote about becoming your own family tree. Assume that an orphan girl grows up, but changes into a man. The man then goes back in time, meets herself, and has a baby girl with her. The man then takes the baby girl further back in time, and drops the baby off at the same orphanage, and then repeats the cycle. In this way, she becomes her own mother, daughter, grandmother, granddaughter, etc.

Ultimately, the final resolution to all these paradoxes may come when the complete theory of quantum gravity is formulated. For example, perhaps when you enter a time machine, your time line might split, and you create a parallel quantum universe. Let's say you go back in time and save Abraham Lincoln from being assassinated at Ford's Theatre. Then perhaps you have saved Abraham Lincoln but in a parallel universe. Hence, the Abraham Lincoln in your original universe did die, and nothing will change that. But the universe has split into two universes, and you have saved President Lincoln in a parallel universe.

So, by assuming the time line can split into a parallel universe, all the paradoxes of time travel can be resolved.

The question of time travel can be definitively answered only when we can calculate the graviton quantum corrections, which we have ignored so far. Physicists have applied the quantum theory to stars and wormholes, but the key is to apply the quantum theory to gravity itself via gravitons, which requires a theory of everything.

This discussion raises interesting questions. Can quantum mechanics fully explain the nature of the Big Bang? Can quantum mechanics applied to gravity answer one of the great questions of science: What happened before the Big Bang?

HOW WAS THE UNIVERSE CREATED?

Where did the universe come from? What set the universe into motion? These are perhaps some of the greatest questions in all of theology and science, the subject of endless speculation.

The ancient Egyptians believed that the universe started as a cosmic egg floating in the Nile. Some Polynesians believed that the universe started with a cosmic coconut. Christians believe the universe was set into motion when God said, "Let there be light!"

The origin of the universe has also fascinated physicists, especially when Newton gave us a compelling theory of gravity. But when Newton tried to apply his theory to the universe we see around us, he encountered problems.

In 1692, he received a disturbing letter from clergyman Richard Bentley. In the letter, Bentley asked Newton to explain a hidden, possibly damaging flaw in his theory. If the universe is finite, and if gravity is always attractive, rather than repulsive, then eventually all the stars in the universe will be attracted to one another. In fact, with enough time, they will all coalesce into a single, gigantic star. So a finite universe should be unstable, and must eventually collapse. Since this does not happen, there must be a flaw in Newton's theory.

Next, he argued that Newton's laws predicted an unstable universe even if the universe were infinite. In an infinite universe, with an infinite number of stars, the sum of all forces tugging on a star from the left and right would also be infinite. Hence, these infinite forces would eventually tear the stars apart, and hence all stars would disintegrate.

Newton was disturbed by this letter, because he had not considered applying his theory to the entire universe. Eventually, Newton came up with a clever but incomplete answer to this question.

Yes, he admitted, if gravity is always attractive, and never repulsive, then the stars in the universe might be unstable. But there was a loophole in this argument. Assume that the universe is, on average, totally uniform and infinite in all directions. In such a static universe, all the forces of gravity cancel one another out, and the universe becomes stable once again. Given any star, the forces of gravity acting on it from all the distant stars in different directions eventually sum to zero, and hence the universe does not collapse.

Although this was a clever solution to this problem,

Newton realized there was still a potential flaw to his solution. The universe might be uniform on average, but it cannot be exactly uniform at all points, so there must be tiny deviations. Like a house of cards, it appears to be stable, but the tiniest flaw will cause the entire structure to collapse. So Newton was clever enough to realize that a uniform infinite universe was indeed stable but was always teetering on the edge of collapse. In other words, the cancellation of infinite forces must be infinitely precise or else the universe will either collapse or be ripped apart.

Thus, Newton's final conclusion was that the universe was infinite and uniform on average, but occasionally God has to tweak the stars in the universe, so they do not collapse under gravity.

WHY IS THE NIGHT SKY BLACK?

But this raised another problem. If we start with a universe that is infinite and uniform, then everywhere we look into space our gaze will eventually hit a star. But since there are an infinite number of stars, there must be an infinite amount of light entering our eyes from all directions.

The night sky should be white, not black. This is called Olbers' paradox.

Some of the greatest minds in history have tried to tackle this sticky question. Kepler, for example, dismissed the paradox by claiming that the universe was finite, and hence there is no paradox. Others have theorized that dust clouds have obscured starlight. (But this cannot explain the paradox, because, in an infinite amount of time, the dust clouds begin to heat up and then emit blackbody radiation, similar to a star. So the universe becomes white again.)

The final answer was actually given by Edgar Allan Poe in 1848. Being an amateur astronomer, he was fascinated by the paradox and said that the night sky is black because, if we travel back in time far enough, we eventually encounter a cutoff—that is, a beginning to the universe. In other words, the night sky is black because the universe has a finite age. We do not receive light from the infinite past, which would make the night sky white, because the universe never had an infinite past. This means that telescopes peering at the farthest stars will eventually reach the blackness of the Big Bang itself.

So it is truly amazing that by pure thought, without doing any experiments whatsoever, one can conclude that the universe must have had a beginning.

GENERAL RELATIVITY AND THE UNIVERSE

Einstein had to confront these puzzling paradoxes when he formulated general relativity in 1915.

Back in the 1920s, when Einstein first began to apply his theory to the universe itself, astronomers told him that the universe was static, neither expanding nor contracting.

But Einstein found something disturbing in his equations. When he tried to solve them, the equations told him that the universe was dynamic, either expanding or contracting. (He did not realize this at the time, but this was the solution to the question asked by Richard Bentley. The universe did not collapse under gravity because the universe was expanding, overcoming the tendency to collapse.)

In order to find a static universe, Einstein was forced to add a fudge factor (called the cosmological constant) into his equations. By adjusting its value by hand, he could cancel out the expansion or contraction of the universe.

Later, in 1929, astronomer Edwin Hubble, by using the giant Mount Wilson Observatory telescope in California, was able to make a startling discovery. The universe was expanding after all, just as Einstein's equations originally predicted. He made this historic discovery by analyzing the Doppler shift of distant galaxies. (When a star moves away from us, the wavelength of its light is stretched so it turns slightly reddish. When the star moves toward us, the wavelength is compressed, so it turns slightly blueish. By carefully analyzing the galaxies, Hubble found that, on average, the galaxies were redshifted and so moving away from us. The universe is expanding.)

In 1931, Einstein visited the Mount Wilson Observatory and met with Hubble. When Einstein was told that the cosmological constant was unnecessary, that the universe was expanding after all, he admitted that the cosmological constant was his "greatest blunder." (Actually, as we shall see, the cosmological constant has made a comeback in recent years, so even his blunders apparently open entirely new areas of scientific investigation.)

It was also possible to take this result one step further and calculate the age of the universe. Since Hubble could calculate the rate at which the galaxies were moving away, it should be possible to "run the videotape backward," and calculate for how long this expansion has taken place. The original answer for the age of the universe came out as 1.8 billion years (which was an embarrassment, since the Earth was known to be older than that—4.6 billion years. But fortunately, the latest satellite data from the Planck satellite puts the age of the universe at 13.8 billion years).

QUANTUM AFTERGLOW OF THE BIG BANG

The next revolution in cosmology took place when physicists began to apply the quantum theory to the Big Bang. Russian physicist George Gamow pondered whether, if the universe started off as a gigantic, superhot explosion, some of that heat would survive today. If we apply the quantum theory to the Big Bang, then the original fireball must have been a quantum blackbody radiator. Since the properties of a blackbody radiator are well-known, it should be possible to calculate the radiation that is the afterglow, or echo, of the Big Bang.

Using the primitive experiments available, in 1948 Gamow and his colleagues Ralph Alpher and Robert Herman calculated that the temperature of the afterglow of the Big Bang should today be around five degrees above absolute zero. (The actual number is 2.73 K.) This is the temperature of the universe after it has cooled for billions of years.

This prediction was verified in 1964 when Arno Penzias and Robert Wilson used the giant Holmdel radio telescope to detect this residual radiation in space. (At first, they thought that this background radiation was due to a defect in their apparatus. According to legend, they realized their mistake when they gave a talk at Princeton, and someone in the audience said, "Either you have detected bird shit, or the creation of the universe." To test this, they had to carefully scrape all the pigeon droppings off the radio telescope.)

Today, this microwave background radiation is perhaps the most persuasive and convincing evidence for the Big Bang. As predicted, recent satellite photographs of the background radiation show a uniform fireball of energy evenly distributed around the universe. (When you hear static on a radio, some of that static actually comes from the Big Bang.)

In fact, these satellite photographs are now so precise that it is possible to detect tiny, minuscule ripples in the background radiation due to the quantum uncertainty principle. At the instant of creation, there should have been quantum fluctuations that caused these ripples. A perfectly smooth Big Bang would have violated the uncertainty principle. These ripples eventually expanded with the Big Bang to create the galaxies we see all around us. (In fact, if our satellites had *not* detected these quantum ripples in the background radiation, their absence would have destroyed the hope of applying the quantum theory to the universe.)

This gives us a remarkable new picture of the quantum theory. The very fact that we exist in the Milky Way galaxy, in the presence of billions of other galaxies, is due to these tiny quantum fluctuations in the original Big Bang. Billions of years ago, everything you see around you was a tiny dot in this background radiation.

The next step in the application of the quantum theory to gravity was when the lessons of the quantum theory and the Standard Model were applied to general relativity.

INFLATION

Buoyed by the success of the Standard Model in the 1970s, physicists Alan Guth and Andrei Linde asked themselves: Could the lessons learned from the Standard Model and the quantum theory be applied directly to the Big Bang?

This was a novel question, since applying the Standard Model to cosmology was still unexplored. Guth noticed that there were two puzzling aspects of the universe that could not be explained by the Big Bang as they conceived of it.

First, there is the flatness problem. Einstein's theory states that the fabric of space-time should have a slight curvature. But when analyzing the curvature of the universe, it seems to be much flatter than predicted by Einstein's theory. In fact, the universe appears to be perfectly flat, to within experimental error.

Second, it is much more uniform than it should be. In the Big Bang, there should have been irregularities and imperfections in the original fireball. Instead, the universe appears to be quite uniform, no matter where we gaze into the heavens.

Both of these paradoxes can be solved by invoking the quantum theory, with a phenomenon Guth called inflation. First, according to this picture, the universe underwent a turbocharged expansion, much faster than originally postulated for the Big Bang. This fantastic expansion basically flattened the universe and eliminated whatever curvature the original universe had.

Second, the original universe might have been irregular, but a tiny piece of that original universe was uniform and was inflated to enormous size. Hence, that would explain why the universe seems to be so uniform today, because we are descended from a tiny, uniform piece of the larger fireball that gave us the Big Bang.

The implications of inflation are far-reaching. It means that the visible universe that we see around us is actually a tiny, infinitesimal piece of a much larger universe, one that we will never see because it is so far away.

But what caused inflation in the first place? What set it in motion? Why did the universe expand at all? Guth then took some inspiration from the Standard Model. In the quantum theory, you start with a symmetry, and then you break it with the Higgs boson to get the universe that we see all around us. Similarly, Guth then theorized that maybe there was a new type of Higgs boson (called the inflaton) that made inflation possible. As with the original Higgs boson, the universe started out in the false vacuum that gave us the era of rapid inflation. But then quantum bubbles occurred within the inflaton field. Inside the bubble, the true vacuum emerged, where the inflation had stopped. Our universe emerged as one of these bubbles. The universe slowed down within the bubble, giving us the present-day expansion.

So far, inflation seems to fit the astronomical data. It is currently the leading theory. But it has unexpected consequences. If we invoke the quantum theory, it means that the Big Bang can happen again and again. New universes may be being born out of our universe all the time.

This means that our universe is actually a single bubble in a bubble bath of universes. This creates a multiverse of parallel universes. This still leaves open a nagging question: What was driving inflation in the first place? That, as we shall see in the next chapter, requires an even more advanced theory, a theory of everything.

RUNAWAY UNIVERSE

General relativity not only gives us unprecedented insight into the beginning of the universe, it also gives us a picture of its ultimate fate. Ancient religions, of course, have given us stark images of the end of time. The ancient Vikings believed the world will end in Ragnarok, or the Twilight of the Gods, when a gigantic snowstorm will engulf the entire planet, and the gods will fight the final battle against their celestial enemies. To Christians, the Book of Revelation foretells disasters, cataclysms, and the coming of the Four Horsemen of the Apocalypse, which precede the Second Coming.

But to a physicist, there are traditionally two ways in which everything will end. If the density of the universe is low, then there is not enough gravity from the stars and galaxies to reverse the cosmic expansion, and the universe will expand forever and slowly die in the Big Freeze. The stars will eventually use up all their nuclear fuel, the sky will turn black, and even black holes will evaporate. The universe will end in a lifeless, supercold sea of drifting subatomic particles.

If the universe is sufficiently dense, then the gravity of

the stars and galaxies might be enough to reverse the cosmic expansion. Then the stars and galaxies will eventually collapse into the Big Crunch, when temperatures soar and devour all life in the universe. (Some physicists have even conjectured that the universe may then bounce back in another Big Bang, creating an oscillating universe.)

But in 1998, astronomers made a stunning announcement that overturned many of our cherished beliefs and forced us to revise our textbooks. By analyzing distant supernovae throughout the universe, they found that the universe was not slowing down in its expansion, as previously thought, but actually speeding up. In fact, it was entering a runaway mode.

They had to revise the two previous scenarios, and a new theory emerged. Perhaps the universe will die in something called the Big Rip, in which the expansion of the universe accelerates to blinding speed. The universe will expand so quickly that the night sky will become totally black (since light cannot reach us from neighboring stars) and everything approaches absolute zero.

At that temperature, life cannot exist. Even the molecules in outer space lose their energy.

What might be driving this runaway expansion is something that was once discarded by Einstein in the 1920s, the cosmological constant, the energy of the vacuum, now called dark energy. Surprisingly, the amount of dark energy in the universe is enormous. More than 68.3 percent of all matter and energy in the universe is in this mysterious form. (Collectively, dark energy and dark matter comprise most of the matter/energy, but they are two distinct entities and should not be confused with each other.)

Ironically, this cannot be explained by any known theory. If one tries to blindly calculate the amount of dark energy in the universe (using the assumptions of relativity and the quantum theory), we find a value that is 10^{120} times larger than the actual value! (That is the number 1 followed by 120 zeros.)

This is the largest mismatch in the entire history of science. The stakes could not be higher: the ultimate fate of the universe itself is hanging in the balance.

This could tell us how the universe itself will die.

WANTED: THE GRAVITON

Although research in general relativity stagnated for decades, the recent application of the quantum to relativity has opened up new unexpected vistas, especially as powerful new instruments go online. There has been a blossoming of new research.

But so far, we have only discussed applying quantum mechanics to the matter that moves within the gravity fields of Einstein's theory. We have not discussed a much more difficult question: applying quantum mechanics to gravity itself in the form of gravitons.

And this is where we encounter the biggest question of all: finding a quantum theory of gravity, which has frustrated the world's great physicists for decades. So let us review what we have learned so far. We recall that when we apply the quantum theory to light, we introduce the photon, a particle of light. As this photon moves, it is surrounded by electric and magnetic fields that oscillate and permeate space and obey Maxwell's equations. This is the reason why light has both particle-like and wavelike properties. The power of Maxwell's equations lies in their symmetries—that is, the ability to turn electric and magnetic fields into each other.

When the photon bumps into electrons, the equation that describes this interaction yields results that are infinite. However, using the bag of tricks devised by Feynman, Schwinger, Tomonaga, and many others, we are able to hide all the infinities. The resulting theory is called QED. Next, we applied this method to the nuclear force. We replaced the original Maxwell field with the Yang-Mills field, and replaced the electron with a series of quarks, neutrinos, etc. Then we introduced a new bag of tricks devised by 't Hooft and his colleagues to eliminate all the infinities once again.

So three of the four forces of the universe could now be unified into a single theory, the Standard Model. The resulting theory was not very pretty, since it was created by cobbling together the symmetries of the strong, weak, and electromagnetic forces, but it worked. But when we apply this tried-and-true method to gravity, we have problems.

In theory, a particle of gravity should be called the graviton. Similar to the photon, it is a point particle, and as it moves at the speed of light, it is surrounded by waves of gravity that obey Einstein's equations.

So far, so good. The problem occurs when the graviton bumps into other gravitons and also atoms. The resulting collision creates infinite answers. When one tries to apply the bag of tricks painfully formulated over the last seventy years, we find that they all fail. The greatest minds of the century have tried to solve this problem, but no one has been successful.

Clearly, an entirely new approach must be used, since all the easy ideas have been investigated and discarded. We need something truly fresh and original. And that leads us to perhaps the most controversial theory in physics, string theory, which might just be crazy enough to be the theory of everything.

6

RISE OF STRING THEORY: PROMISE AND PROBLEMS

We saw earlier that around 1900, there were two great pillars of physics: Newton's law of gravity and Maxwell's equations for light. Einstein realized that these two great pillars were in conflict with each other. One of them would have to collapse. The fall of Newtonian mechanics set into motion the great scientific revolutions of the twentieth century.

Today, history may be repeating itself. Once again we have two great pillars of physics. On one hand, we have the theory of the very big, Einstein's theory of gravity, which gives us black holes, the Big Bang, and the expanding universe. On the other hand, we have the theory of the very small, the quantum theory, which explains the behavior of subatomic particles. The problem is that they stand in conflict with each other. They are based on two different principles, two different mathematics, and two different philosophies.

The next great revolution, we hope, will be to unify these two pillars into one.

STRING THEORY

It all began in 1968, when two young physicists, Gabriele Veneziano and Mahiko Suzuki, were thumbing through math books and stumbled across a strange formula found by mathematician Leonhard Euler in the eighteenth century. This strange formula seemed to describe the scattering of two subatomic particles! How could an abstract formula from the eighteenth century describe the latest results from our atom smashers? Physics was not supposed to work this way.

Later, physicists, including Yoichiro Nambu, Holger Nielsen, and Leonard Susskind, realized that the properties of this formula represented the interaction of two strings. Very quickly, this formula was generalized to a whole army of equations, representing the scattering of multistrings. (This was, in fact, my Ph.D. thesis, calculating the complete set of interactions of an arbitrary number of strings.) Then researchers were able to introduce spinning particles into string theory.

String theory was like an oil well suddenly gushing forth a torrent of new equations. (Personally, I was not satisfied with this, because, ever since Faraday, physics had been represented by fields that concisely summarized vast amounts of information. String theory, by contrast, was a collection of disjointed equations. My colleague Keiji Kikkawa and I were then successful in writing all of string theory in the language of fields, creating what is called string field theory. All of string theory can be summarized by our equations in a field theory equation just one inch long.)

As a result of the torrent of equations, a new picture was beginning to emerge. Why were there so many particles? Like Pythagoras more than two thousand years ago, the theory said that each musical note—each vibration of a string—represented a particle. Electrons, quarks, and Yang-Mills particles were nothing but different notes on the same vibrating string.

What is so powerful and interesting about the theory is that gravity is necessarily included. Without any extra assumptions, the graviton emerges as one of the lowest vibrations of the string. In fact, even if Einstein had never been born, his entire theory of gravity might have been found simply by looking at the lowest vibration of the string.

As physicist Edward Witten once said, "String theory is extremely attractive because gravity is forced upon us. All known consistent string theories include gravity, so while gravity is impossible in quantum field theory as we have known it, it's obligatory in string theory."

TEN DIMENSIONS

But as the theory began to evolve, more and more fantastic, totally unexpected features began to be revealed. For example, it was found that the theory can only exist in ten dimensions!

This shocked physicists, because no one had ever seen anything like it. Usually, any theory can be expressed in any dimension you like. We simply discard these other theories because we obviously live in a three-dimensional world. (We can only move forward, sideways, and up and down. If we add time, then it takes four dimensions to locate any event in the universe. If we want to meet someone in Manhattan, for example, we might say, Let's meet at the corner of 5th Avenue and 42nd Street, on the tenth floor, at noon. However, moving in dimensions beyond four is impossible for us, no matter how we try. In fact, our brains cannot even visualize how to move in higher dimensions. Therefore all the research done in higher-dimensional string theory is done using pure mathematics.)

But in string theory, the dimensionality of space-time is fixed at ten dimensions. The theory breaks down mathematically in other dimensions.

I still remember the shock that physicists felt when string theory posited that we live in a universe of ten dimensions. Most physicists saw this as proof that the theory was wrong. When John Schwarz, one of the leading architects of string theory, was in the elevator at Caltech, Richard Feynman would prod him, asking, "Well, John, and how many dimensions are you in today?"

Yet over the years, physicists gradually began to show that all rival theories suffered from fatal flaws. For example, many could be ruled out because their quantum corrections were infinite or anomalous (that is, mathematically inconsistent).

So over time, physicists began to warm up to the idea that perhaps our universe might be ten-dimensional after all. Finally, in 1984, John Schwarz and Michael Green showed that string theory was free of all the problems that had doomed previous candidates for a unified field theory.

If string theory is correct, then the universe might have originally been ten-dimensional. But the universe was unstable and six of these dimensions somehow curled up and became too small to be observed. Hence, our universe might actually be ten-dimensional, but our atoms are too big to enter these tiny higher dimensions.

THE GRAVITON

In spite of all the craziness of string theory, one thing that has kept it alive is that it successfully marries the two great theories of physics, general relativity and the quantum theory, giving us a finite theory of quantum gravity. That is what all the excitement is about.

Previously, we mentioned that if you add quantum corrections to QED, or the Yang-Mills particle, you get a flood of infinities that must be carefully and tediously removed.

But all this fails when we try to have a shotgun wedding between the two great theories of nature, relativity and the quantum theory. When we apply the quantum principle to gravity, we have to break it up into packets of energy, or quanta, called the graviton. Then we calculate the collision of these gravitons with other gravitons and with matter, like the electron. But when we do this, the entire bag of tricks found by Feynman and 't Hooft fail miserably. The quantum corrections caused by gravitons interacting with other gravitons are infinite and defy all the methods found by previous generations of physicists.

This is where the next magic occurs. String theory can remove these troublesome infinites that have dogged physicists for almost a century. And this magic once again occurs through symmetry.

SUPERSYMMETRY

Historically, it was always considered nice to have our equations symmetrical, but it was a luxury that was not strictly necessary. But in the quantum theory, symmetry becomes the most important feature of the physics.

As we've established, when we calculate the quantum corrections to a theory, these quantum corrections are often divergent (that is, infinite), or anomalous (meaning that it violates the original symmetry of the theory). Physicists have realized only in the last few decades that symmetry, instead of being just a pleasing feature of a theory, is actually the central ingredient. *Demanding a theory be symmetrical can often banish the divergences and anomalies that plague nonsymmetrical theories.* Symmetry is the sword physicists use to vanquish the dragons unleashed by quantum corrections.

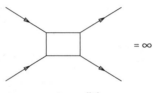

two gravitons collide

two strings collide

Figure 11. When calculating the collision of two gravitons (*top*), the answer is infinite and hence meaningless. But when two strings collide (*bottom*), we have two terms, one from the bosons and one from the fermions. In string theory, these two terms cancel exactly, helping to create a finite theory of quantum gravity.

As we mentioned earlier, Dirac found that his equation for the electron predicted that it had spin (which is a mathematical feature of the equations that resembles the familiar spin we see all around us). Later, physicists found that all subatomic particles have spin. But spin comes in two types.

In certain quantum units, the spin can be either integral (like 0, 1, or 2) or half-integral (like ½, ³⁄₂). First, the particles that have integral spin describe the forces of the universe. They include the photon and Yang-Mills particle (with spin 1) and the particle of gravity, the graviton (with spin 2). These particles are named bosons (after the Indian physicist Satyendra Nath Bose). So the forces of nature are mediated by bosons.

Then there are particles that make up the matter in the universe. They have half-integral spin, such as electrons, neutrinos, and quarks (with spin ½). These particles are called fermions (after Enrico Fermi), out of which we can build up the other particles of the atom: protons and neutrons. So the atoms of our body are collections of fermions.

TWO TYPES OF SUBATOMIC PARTICLES

FERMIONS (MATTER)	BOSONS (FORCES)
electron, quark,	photon, graviton,
neutrino, proton	Yang-Mills

Bunji Sakita and Jean-Loup Gervais then demonstrated that string theory had a new type of symmetry, called supersymmetry. Since then, supersymmetry has been expanded so that it is now the largest symmetry ever found in physics. As we have emphasized, beauty to a physicist is symmetry, which allows us to find the link between different particles. All the particles of the universe could then be unified by supersymmetry. As we have emphasized, a symmetry rearranges the components of an object, leaving the original object the same. Here, one is rearranging the particles in our equations so that fermions are interchanged with bosons and vice versa. This becomes the central feature of string theory, so that the particles of the entire universe can be rearranged into one another.

This means that each particle has a super partner, called a sparticle, or super particle. For example, the super partner of the electron is called the selectron. The super partner of the quark is called the squark. The superpartner of the lepton (like the electron or neutrino) is called the slepton.

But in string theory, something remarkable happens. When calculating quantum corrections to string theory, you have two separate contributions. You have quantum corrections coming from fermions and also bosons. Miraculously, they are equal in size, but occur with the opposite sign. One term might have a positive sign, but there is another term that is negative. In fact, when they are added together, *these terms cancel against each other, leaving a finite result.*

The marriage between relativity and the quantum theory has dogged physicists for almost a century, but the symmetry between fermions and bosons, called supersymmetry, allows us to cancel many of these infinities against each other. Soon, physicists discovered other means of eliminating these infinities, leaving a finite result. So this is the origin of all the excitement surrounding string theory: it can unify gravity with the quantum theory. No other theory can make this claim. This may satisfy Dirac's original objection. He hated renormalization theory because, in spite of its fantastic and undeniable successes, it involved adding and subtracting quantities that were infinite in size. Here, we see that string theory is finite all by itself, without renormalization.

This, in turn, may satisfy the picture originally proposed by Einstein himself. He once compared his theory of gravity to marble, which is smooth, elegant, polished. However, matter, by contrast, was more like wood. The trunk of a tree is gnarled, chaotic, rough, without a regular geometric pattern. His goal was to ultimately create a unified theory that combined the marble and the wood into a single form— that is, *to create a theory entirely made of marble*. That was Einstein's dream.

String theory can complete this picture. Supersymmetry is a symmetry that can turn marble into wood and vice versa. They become two sides of the same coin. In this picture, marble is represented by bosons, and wood is represented by fermions. Although there is no experimental evidence for supersymmetry in nature, it is so elegant and beautiful that it has captured the imagination of the physics community.

As Steven Weinberg once said, "Although the symmetries are hidden from us, we can sense that they are latent in nature, governing everything about us. That's the most exciting idea I know: that nature is much simpler than it looks. Nothing makes me more hopeful that our generation of human beings may actually hold the key to the universe in our hands—that perhaps in our lifetimes we may be able to tell why all of what we see in this immense universe of galaxies and particles is logically inevitable."

In summary, we now see that symmetry may be the key to unifying all the laws of the universe, due to several remarkable achievements:

· Symmetry creates order out of disorder. Out of the chaos of chemical elements and subatomic particles, the Mendeleyev periodic table and Standard Model can rearrange them in a tidy, symmetric fashion.
· Symmetry helps fill in the gaps. Symmetry allows you to isolate gaps in these theories and hence predict the existence of new types of elements and subatomic particles.
· Symmetry unifies totally unexpected and seemingly unrelated objects. Symmetry finds the link between space and time, matter and energy, electricity and magnetism, and fermions and bosons.

· Symmetry reveals unexpected phenomena. Symmetry predicted the existence of new phenomena such as antimatter, spin, and quarks.

· Symmetry eliminates unwanted consequences that can destroy the theory. Quantum corrections often have disastrous divergences and anomalies that can be eliminated by symmetry.

· Symmetry alters the original classical theory. The quantum corrections to string theory are so stringent they actually alter the original theory, fixing the dimensionality of space-time.

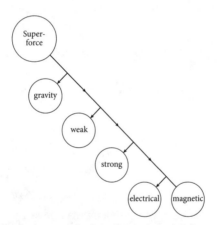

Figure 12. At the beginning of time, it is believed there was a single superforce whose symmetry included all the particles of the universe. But it was unstable, and the symmetry began to break. The first to split off was gravity. Then the strong force and the weak force followed, leaving the electromagnetic force. So the universe today looks broken, with all the forces quite different from one another. It is the job of physicists to reassemble the pieces back together into a single force.

Superstring theory takes advantage of all these features. Its symmetry is supersymmetry (the symmetry that can interchange bosons and fermions). Supersymmetry, in turn, is the largest symmetry ever found in physics, capable of unifying all the known particles of the universe.

M-THEORY

We have yet to complete the last step in string theory, finding its fundamental physical principles—that is, we still don't understand how to derive the entire theory from a single equation. One shock wave came in 1995, when string theory underwent another metamorphosis and a new theory emerged, called M-theory. The problem with the original string theory was that there were five distinct versions of quantum gravity, each of them finite and well defined. These five string theories looked very similar, except their spins were arranged slightly differently. People began to ask: Why should there be five? Most physicists thought that the universe should be unique.

Physicist Edward Witten found that there was actually a hidden eleven-dimensional theory, called M-theory, that was based on membranes (like the surfaces of spheres and doughnuts) rather than just strings. He was able to explain why there were five different string theories, because there were five ways in which to collapse an eleven-dimensional membrane to a ten-dimensional string.

In other words, all five versions of string theory were different mathematical representations of the same M-theory. (So string theory and M-theory are really the same theory, except that string theory is a reduction of eleven-dimensional M-theory to ten dimensions.) But how can a single eleven-dimensional theory give rise to five ten-dimensional theories?

For example, think of a beach ball. If we let the air out, the ball collapses, gradually resembling a sausage. If we let even more air out, the sausage becomes a string. Hence, a string is actually a membrane in disguise, such that its air has been let out.

If we start with a eleven-dimensional beach ball, you can show mathematically that there are five ways in which it can be collapsed to a ten-dimensional string.

Or think of the tale of the blind men who encounter an elephant for the first time. One wise man, touching the ear of the elephant, declares the elephant is flat and two-dimensional like a fan. Another wise man touches the tail and assumes the elephant is like rope or a one-dimensional string. Another, touching a leg, concludes the elephant is a three-dimensional drum or a cylinder. But actually, if we step back and rise into the third dimension, we can see the elephant as a three-dimensional animal. In the same way, the five different string theories are like the ear, tail, and leg, but we still have yet to reveal the full elephant, M-theory.

HOLOGRAPHIC UNIVERSE

As we mentioned, with time new layers have been uncovered in string theory. Soon after M-theory was proposed in 1995, another astonishing discovery was made by Juan Maldacena in 1997.

He jolted the entire physics community by showing something that was once considered impossible: that a supersymmetric Yang-Mills theory, which describes the behavior of subatomic particles in four dimensions, was dual, or mathematically equivalent, to a certain string theory in ten dimensions. This sent the physics world into a tizzy. By 2015, there were ten thousand papers that referred to this paper, making it by far the most influential paper in high-energy physics. (Symmetry and duality are related but different. Symmetry arises when we rearrange the components of a single equation and it remains the same. Duality arises when we show that two entirely different theories are actually mathematically equivalent. Remarkably, string theory has both of these highly nontrivial features.)

As we saw, Maxwell's equations have a duality between electric and magnetic fields—that is, the equations remain the same if we reverse the two fields, turning electric fields into magnetic fields. (We can see this mathematically, because the EM equations often contain terms like $E^2 + B^2$, which remain the same when we rotate the two fields into each other, like in the Pythagorean theorem). Similarly, there are five distinct string theories in ten dimensions, which can be proven to be dual to each other, so they are really a single eleven-dimensional M-theory in disguise. So remarkably, duality shows that two different theories are actually two aspects of the same theory.

Maldacena, however, showed that there was yet another duality between strings in ten dimensions and Yang-Mills theory in four dimensions. This was a totally unexpected development but one that has profound implications. It meant that there were deep, unexpected connections between the gravitational force and the nuclear force defined in totally different dimensions.

Usually, dualities can be found between strings in the same dimension. By rearranging the terms describing those strings, for example, we can often change one string theory into another. This creates a web of dualities between different string theories, all defined in the same dimension. But a duality between two objects defined in different dimensions was unheard of.

This is not an academic question, because it has far-reaching implications for understanding the nuclear force. For example, earlier we saw how gauge theory in four dimensions, as represented by the Yang-Mills field, gives us the best description of the nuclear force, but no one has ever been able to find an exact solution to the Yang-Mills field. But since gauge theory in four dimensions could be dual to string theory in ten dimensions, it meant that quantum gravity might hold the key to the nuclear force. This was an astonishing revelation, because it meant that basic features of the nuclear force (such as calculating the mass of the proton) might be best described by string theory.

This created a bit of an identity crisis among physicists. Those who work exclusively on the nuclear force spend all their time studying three-dimensional objects, such as protons and neutrons, and often scoff at physicists theorizing in higher dimensions. But with this new duality between gravity and gauge theory, suddenly these physicists found themselves trying to learn all about ten-dimensional string theory, which might hold the key to understanding the nuclear force in four dimensions.

Yet another unexpected development emerged from this bizarre duality, called the holographic principle. Holograms are two-dimensional flat sheets of plastic, containing the image of three-dimensional objects that have been specially encoded within them. By shining a laser beam at the flat screen, the three-dimensional image suddenly emerges. In other words, all the information needed to create a three-dimensional image has been encoded onto a flat two-dimensional screen using lasers, like the image of Princess Leia projected by R2-D2 or the haunted mansion at Disneyland where three-dimensional ghosts sail around us.

This principle also works for black holes. As we saw earlier, if we throw an encyclopedia into a black hole, the information contained inside the books cannot disappear, according to quantum mechanics. So where does the information go? One theory posits that it is distributed onto the surface of the event horizon of the black hole. So the two-dimensional surface of a black hole contains all the information of all the three-dimensional objects that have been thrown into it.

This also has implications for our conception of reality. We are convinced, of course, that we are three-dimensional objects that can move in space, defined by three numbers, length, width, and height. But perhaps this is an illusion. Perhaps we are living in a hologram.

Perhaps the three-dimensional world we experience is just a shadow of the real world, which is actually ten- or eleven-dimensional. When we move in the three dimensions of space, we experience our real selves actually moving in ten or eleven dimensions. When we walk down the street, our shadow follows us and moves like us, except the shadow exists in two dimensions. Likewise, perhaps we are shadows moving in three dimensions, but our real selves are moving in ten or eleven dimensions.

In summary, we see that with time, string theory reveals new, totally unexpected results. It means that we still do not really understand the basic fundamental principles behind it. Eventually, it may turn out that string theory is not really a theory about strings after all, since strings can be expressed as membranes when formulated in eleven dimensions.

That is why it is premature to compare string theory with experiment. Once we have revealed the true principles behind string theory, we may find a way to test it, and maybe then we can say once and for all if it is a theory of everything or a theory of nothing.

TESTING THE THEORY

But despite all the theoretical successes of string theory, it still has glaring weak spots. Any theory that makes claims as powerful as the ones made for string theory is naturally going to attract an army of detractors. One has to be continually reminded of the words of Carl Sagan, who said that "remarkable claims require remarkable proof."

(I am also reminded of the cynical words of Wolfgang Pauli, who was a master of the put-down. When listening to a talk, he might say, "What you said was so confused that one could not tell whether it was nonsense or not." He would also say, "I do not mind if you think slowly, but I do object when you publish more quickly than you think." If he were alive, he might apply these words to string theory.)

The debate is so intense that the best minds in physics have split on this question. Not since the great sixth Solvay Conference of 1930, when Einstein and Bohr sparred with each other on the question of the quantum theory, has science witnessed such a grand schism.

Nobel laureates have taken opposite positions on this question. Sheldon Glashow has written, "Years of intense effort by dozens of the best and the brightest have yielded not one verifiable prediction, nor should any soon be expected." Gerard 't Hooft went so far as to say that the interest surrounding string theory is comparable to "American television commercials"—that is, all hype and fanfare, but no substance.

Others have praised the virtues of string theory. David Gross has written, "Einstein would have been pleased with this, at least with the goal, if not the realization. . . . He would have liked the fact that there is an underlying geometrical principle—which, unfortunately, we don't really yet understand."

Steven Weinberg has compared string theory to the historic effort to find the north pole. All ancient maps of the Earth had a huge, gaping hole, where the north pole should be, but no one had ever seen it. Anywhere on the Earth, all compass needles pointed to this mythical place. But all attempts to find the fabled north pole ended in failure. In their hearts, the ancient mariners knew that there must be a north pole, but no one could prove it. Some even doubted that it existed. However, after centuries of speculation, finally in 1909 Robert Peary actually set foot on the north pole.

String theory critic Glashow has admitted that he is outnumbered in this debate. He once commented, "I find myself a dinosaur in a world of upstart mammals."

CRITICISMS OF STRING THEORY

There are several main criticisms that have been leveled at string theory. The critics have claimed that the theory is all hype; that beauty by itself is an unreliable guide in physics; that it predicts too many universes; and, most important, that it is untestable.

The great astronomer Kepler was once misled by the power of beauty. He was enamored of the fact that the solar system resembled a collection of regular polyhedrons stacked inside one another. Centuries earlier, the Greeks had enumerated five of these polyhedrons (e.g., the cube, pyramid, etc.). Kepler noticed that by sequentially putting these polyhedrons inside one another, like Russian dolls, one could reproduce some of the details of the solar system. It was a beautiful idea, but turned out to be totally wrong.

Recently, some physicists have criticized string theory, stating that beauty is a misleading criterion for physics. Just because string theory has brilliant mathematical properties does not mean it holds a kernel of truth. They rightly point out that beautiful theories have sometimes been dead ends.

But poets often quote the poem "Ode on a Grecian Urn" by John Keats:

Beauty is truth, truth beauty,—that is all
Ye know on earth, and all ye need to know.

Paul Dirac was certainly a follower of this principle when he wrote, "The research worker, in his efforts to express the fundamental laws of Nature in mathematical form, should strive mainly for mathematical beauty." In fact, he would write that he discovered his celebrated theory of the electron by fiddling with pure mathematical formulas rather than looking at the data.

As powerful as beauty is in physics, certainly beauty can often lead you astray. As physicist Sabine Hossenfelder has written, "Beautiful theories have been ruled out in the hundreds, theories about unified forces and new particles and additional symmetries and other universes. All these theories were wrong, wrong, wrong. Relying on beauty is clearly not a successful strategy."

The critics claim string theory has beautiful mathematics, but this may have nothing to do with physical reality.

There is some validity to this criticism, but one has to realize that aspects of string theory like supersymmetry are not useless and devoid of physical applications. Although evidence for supersymmetry has not yet been found, it has proven to be essential in eliminating many of the defects within the quantum theory. Supersymmetry, by canceling bosons against fermions, enables us to solve a long-standing problem, eliminating the divergences that plague quantum gravity.

Not every beautiful theory has a physical application, but all fundamental physical theories found so far, without exception, have a type of beauty or symmetry built into them.

CAN IT BE TESTED?

The foremost criticism of string theory is that it is untestable. The energy that gravitons possess is called the Planck energy, which is a quadrillion times greater than the energy produced by the LHC. Imagine trying to build a LHC that is a quadrillion times larger than the current one! One would probably need a particle accelerator the size of the galaxy for a direct test of the theory.

Furthermore, each solution of string theory is an entire universe. And there seems to be an infinite number of solutions. For a direct test of the theory, one would need to create baby universes in the laboratory! In other words, only a god can truly test the theory directly, since the theory is based on universes, not just atoms or molecules.

So at first, it seems that string theory fails the acid test for any theory, testability. But promoters of string theory are not fazed. As we have established, most science is done indirectly, by examining echoes from the sun, the Big Bang, etc.

Similarly, we look for echoes from the tenth and eleventh dimension. Perhaps evidence for string theory is hidden all around us, but we have to listen for its echoes, rather than try to observe it directly.

For example, one possible signal from hyperspace is the existence of dark matter. Until recently, it was widely believed that the universe is mainly made of atoms. Astronomers have been shocked to find that only 4.9 percent of the universe is made of atoms like hydrogen and helium. Actually, most of the universe is hidden from us, in the form of dark matter and dark energy. (We recall that dark matter and dark energy are two distinct things. Twenty-six point eight percent of the universe is made of dark matter, which is invisible matter that surrounds the galaxies and keep them from flying apart. And 68.3 percent of the universe is made of dark energy, which is even more mysterious, the energy of empty space that is driving the galaxies apart.) Perhaps evidence for the theory of everything lies hidden in this invisible universe.

SEARCH FOR DARK MATTER

Dark matter is strange, it is invisible, yet it holds the Milky Way galaxy together. But since it has weight and no charge, if you tried to hold dark matter in your hand it would sift through your fingers as if they weren't there. It would fall right through the floor, through the core of the Earth, and then to the other side of the Earth, where gravity would eventually cause it to reverse course and fall back to your location. It would then oscillate between you and the other side of the planet, as if the Earth weren't there.

As strange as dark matter is, we know it must exist. If we analyze the spin of the Milky Way galaxy and use Newton's laws, we find that there is not enough mass to counteract the centrifugal force. Given the amount of mass we see, the galaxies in the universe should be unstable and they should fly apart, but they have been stable for billions of years. So we have two choices: either Newton's equations are incorrect when applied to galaxies, or else there is an unseen object that is keeping the galaxies intact. (We recall that the planet Neptune was found in the same way, by postulating a new planet that explained Uranus's deviations from a perfect ellipse.)

At present, one leading candidate for dark matter is called the weakly interacting massive particles (WIMPs). Among them, one likely possibility is the photino, the supersymmetric partner of the photon. The photino is stable, has mass, is invisible, and has no charge, which fits precisely the characteristics of dark matter. Physicists believe the Earth moves in an invisible wind of dark matter that is probably passing through your body right now. If a photino collides with a proton, it may cause the proton to shatter into a shower of subatomic particles that can then be detected. In fact, even today there are huge swimming pool–sized detectors (with vast amounts of fluids containing xenon and argon) that may one day capture the spark created by a photino collision. There are about twenty active groups searching for dark matter, often deep inside mine shafts below the Earth's surface, away from interfering cosmic ray interactions. So it is conceivable that the collision of dark matter may be captured by our instruments. Once dark matter collisions have been detected, then physicists will study the properties of dark matter particles and then compare them to the predicted properties of photinos. If the predictions of string theory match the experimental results on dark matter, this would go a long way toward convincing physicists that this is the correct path.

Another possibility is that the photino may be produced by the next generation of particle accelerators being discussed.

BEYOND THE LHC

The Japanese are considering funding the International Linear Collider, which would shoot a beam of electrons down a straight tube, until it strikes a beam of anti-electrons. If approved, the device would be built in twelve years. The advantage of a collider like this is that it uses electrons rather than protons. Because protons consist of three quarks held together by gluons, the collision between protons is very messy, with an avalanche of extraneous particles being created. The electron, by contrast, is a single elementary particle, so the collision with an anti-electron is much cleaner and requires much less energy. As a result, at only 250 billion electron volts, it should be able to create Higgs bosons.

The Chinese have also expressed interest in building the Circular Electron Positron Collider. Work would begin around 2022, and it might be finished around 2030, at a cost of $5 to $6 billion. It would reach an energy of 240 billion electron volts and would be 100 kilometers around.

Not to be outdone, the physicists at CERN are planning the successor to the LHC, called the Future Circular Collider (FCC). It would eventually reach an astounding 100 trillion electron volts. It would also be about 100 kilometers around.

It is not clear if these accelerators will ever be built, but it does mean there is hope for finding dark matter in the next generation of accelerators beyond the LHC. If we discover particles of dark matter, they can then be compared against the predictions of string theory.

Another prediction of string theory that might be verified by these accelerators is the presence of mini black holes.

Since string theory is a theory of everything, it includes gravity as well as subatomic particles, so physicists expect to find tiny black holes in the accelerator. (These mini black holes, unlike stellar black holes, are harmless and have the energy of tiny subatomic particles, not the energy of dying stars. In fact, the Earth is bombarded by cosmic rays much more powerful than any that can be produced by these accelerators, without any harmful effects.)

BIG BANG AS ATOM SMASHER

There is also the hope that we can take advantage of the greatest atom smasher of all, the Big Bang itself. Radiation from the Big Bang may give us a clue to dark matter and dark energy. First of all, the echo, or afterglow, of the Big Bang is easy to detect. Our satellites have been able to detect this radiation to enormous accuracy.

Photographs of this microwave background radiation show that it is remarkably smooth, with tiny ripples appearing on its surface. These ripples, in turn, represent tiny quantum fluctuations that existed at the instant of the Big Bang that were then magnified by the explosion.

What is controversial, however, is that there appear to be irregularities, or blotches, in the background radiation that we cannot explain. There is some speculation that these strange blotches are the remnants of collisions with other universes. In particular, the CMB (cosmic microwave background) cold spot is an unusually cool mark on the otherwise uniform background radiation that some physicists have speculated might be the remnants of some type of connection or collision between our universe and a parallel universe at the beginning of time. If these strange markings represent our universe interacting with parallel universes, then the multiverse theory might become more plausible to skeptics.

Already, there are plans to put detectors in space that can refine all these calculations, using space-based gravity wave detectors.

LISA

Back in 1916, Einstein showed that gravity could travel in waves. Like throwing a stone in a pond and witnessing the concentric, expanding rings it creates, Einstein predicted that swells of gravity would travel at the speed of light. Unfortunately, these would be so faint that he did not think we would find them anytime soon.

He was right. It took until 2016, one hundred years after his original prediction, before gravity waves were observed. Signals from two black holes that collided in space about a billion years ago were captured by huge detectors. These detectors, built in Louisiana and Washington State, each occupy several square miles of real estate. They resemble a large *L,* with laser beams traveling down each leg of the *L*. When the two beams meet at the center, they create an interference pattern that is so sensitive to vibrations that they could detect this collision.

For their pioneering work, three physicists, Rainer Weiss, Kip S. Thorne, and Barry C. Barish, won the Nobel Prize in 2017.

For even greater sensitivity, there are plans to send gravity wave detectors into outer space. The project, known as the laser interferometry space antenna (LISA), might be able to pick up vibrations from the instant of the Big Bang itself. One version of the LISA consists of three separate satellites in space, each connected to the others by a network of laser beams. The triangle is about a million miles on each side. When a gravity wave from the Big Bang hits the detector, it causes the laser beams to jiggle a bit, which can then be measured by sensitive instruments.

The ultimate goal is to record the shock waves from the Big Bang, and then run the videotape backward to get the best guess for the radiation before the Big Bang. These pre–Big Bang waves would then be compared to what's predicted in several versions of string theory. In this way, one might be able to get numerical data about the multiverse before the Big Bang.

Using devices more advanced than LISA, one might be able to get baby pictures of the universe. And perhaps even find evidence of the umbilical cord connecting our infant universe to a parent universe.

TESTING THE INVERSE SQUARE LAW

Another frequent objection to string theory is that it postulates that we actually live in ten or eleven dimensions, for which there is no experimental evidence.

But this aspect might actually be testable with off-the-shelf instruments. If our universe is three-dimensional, then the force of gravity diminishes as the square of the distance of separation. This famous law of Newton is what guides our space probes millions of miles in space with breathtaking precision, so we can shoot space probes right through the rings of Saturn if we felt like it. But Newton's famous inverse square law has been tested only over astronomical distances, rarely in the laboratory. If the strength of gravity over small distances does not obey the inverse square law, it would signal the presence of a higher dimension. For example, if the universe had four spatial dimensions, then gravity should diminish as the cube of the distance of separation. (If the universe had N spatial dimensions, then gravity should diminish with the $(N-1)$ power of the distance of separation.)

But rarely has the force of gravity been measured between two objects in the laboratory. These experiments are difficult to do, since gravitational forces are quite small in the laboratory, but the first measurements have been done in Colorado, and the results were negative—that is, Newton's inverse square law still holds. (But this means only that there are no added dimensions in Colorado.)

LANDSCAPE PROBLEM

To a theoretician, all these criticisms are troublesome but not fatal. But what does cause problems for a theoretician is that the model seems to predict a multiverse of parallel universes, many of which are crazier than those in the imagination of a Hollywood scriptwriter. String theory has an infinite number of solutions, each describing a perfectly well-behaved finite theory of gravity, which do not resemble our universe at all. In many of these parallel universes, the proton is not stable, so it would decay into a vast cloud of electrons and neutrinos. In these universes, complex matter as we know it (atoms and molecules) cannot exist. They only consist of a gas of subatomic particles. (Some might argue that these alternate universes are only mathematical possibilities and are not real. But the problem is that the theory lacks predictive power, since it cannot tell you which of these alternate universes is the real one.)

This problem is actually not unique to string theory. For example, how many solutions are there to Newton's or Maxwell's equations? There are an infinite number, depending on what you are studying. If you start with a light bulb or a laser and you solve Maxwell's equations, you find a unique solution for each instrument. So Maxwell's or Newton's theories also have an infinite number of solutions, depending on the initial conditions—that is, the situation you start with.

This problem is likely to exist for *any* theory of everything. Any theory of everything will have an infinite number of solutions depending on the initial conditions. But how do you determine the initial conditions of the entire universe? This means you have to input the conditions of the Big Bang from the outside, by hand.

To many physicists this seems like cheating. Ideally, you want the theory itself to tell you the conditions that gave rise to the Big Bang. You want the theory to tell you everything, including the temperature, density, and composition of the original Big Bang. A theory of everything should somehow contain its own initial conditions, all by itself.

In other words, you want a unique prediction for the beginning of the universe. So string theory has an embarrassment of riches. Can it predict our universe? Yes. That is a sensational claim, the goal of physicists for almost a century. But can it predict just one universe? Probably not. This is called the landscape problem.

There are several possible solutions to this problem, none of them widely accepted. The first is the anthropic principle, which says that our universe is special because we, as conscious beings, are here to discuss this question in the first place. In other words, there might be an infinite number of universes, but our universe is the one that has the conditions that make intelligent life possible. The initial conditions of the Big Bang are fixed at the beginning of time so that intelligent life can exist today. The other universes might have no conscious life in them.

I clearly remember my first introduction to this concept when I was in the second grade. I remember my teacher said that God so loved the Earth that he put the Earth "just right" from the sun. Not too close, or the oceans would boil. Not too far, or the oceans would freeze. Even as a child, I was stunned by this argument, because it used pure logic to determine the nature of the universe. But today, satellites have revealed four thousand planets orbiting other stars. Sadly, most of them are too close or too far from their star to support life. So there are two ways one can analyze my second-grade teacher's argument. Perhaps there is a loving God after all, or perhaps there are thousands of dead planets that are too close or too far, and we are on a planet that is just right for sustaining intelligent life that hence can debate this question. Similarly, we may coexist in an ocean of dead universes, and our universe is special only because we are here to discuss this question.

The anthropic principle actually allows one to explain a curious experimental fact about our universe: that the basic constants of nature seem to be fine-tuned to allow for life.

As physicist Freeman Dyson has written, it seems as if the universe knew that we were coming. For example, if the nuclear force were a bit weaker, the sun would never have ignited, and the solar system would be dark. If the strong nuclear force were a bit stronger, then the sun would have burned out billions of years ago. So the nuclear force is tuned just right.

Similarly, if gravity were a bit weaker, perhaps the Big Bang would have ended in a Big Freeze, with a dead, cold expanding universe. If gravity were a bit stronger, we might have ended in a Big Crunch, and all life would have been burned to death. Yet our gravity is just right to allow for stars and planets to form and last long enough for life to spring up.

One can list a number of these accidents that make life possible, and each time we are in the middle of the Goldilocks zone. So the universe is one gigantic crapshoot, and we won the roll. But according to the multiverse theory, it means we coexist with a vast number of dead universes.

So perhaps the anthropic principle can pick our universe from the millions of universes in the landscape, because we have conscious life in our universe.

MY OWN POINT OF VIEW ON STRING THEORY

I have been working on string theory since 1968, so I have my own definite viewpoint. However you look at it, the final form of the theory has yet to be revealed. So it is premature to compare string theory to the present universe.

One feature of string theory is that it is evolving backward, revealing new mathematics and concepts along the way. Every decade or so, there is a new revelation in string theory that changes our point of view concerning its nature. I have witnessed about three such astonishing revolutions, yet we have yet to express string theory in its complete form. We do not yet know its final fundamental principles. Only then can we compare it with experiment.

REVEALING A PYRAMID

I like to compare it to searching for treasure in the Egyptian desert. Let's say one day you stumble on a tiny rock sticking up in the desert. After brushing away the sand, you begin to realize that this pebble is actually the top of a gigantic pyramid. After years of excavation, you find all sorts of strange chambers and artwork. In each floor, you find new surprises. Finally, after excavating many floors, you reach the final door, and are about to open it to find out who made the pyramid.

Personally, I believe we are still not at the bottom floor, since we keep discovering new mathematical layers every time we analyze the theory. There are still more layers to reveal before we find string theory in its final form. In other words, the theory is smarter than we are.

It is possible to express all of string theory in terms of string field theory in an equation about one inch long. But we need five such equations in ten dimensions.

Although we can express string theory in field theory form, this is still not possible for M-theory. The hope is that one day physicists may find a single equation that summarizes all of M-theory. Unfortunately, it is notoriously difficult to express a membrane (which can vibrate in so many ways) in field theory form. As a consequence, M-theory consists of scores of disjointed equations that miraculously describe the same theory. If we can write M-theory in field theory form, then the entire theory should emerge from a single equation.

No one can predict if or when this will happen. But after witnessing the hype around string theory, the public has grown impatient.

But even among string theorists, there is a certain amount of pessimism about the future prospects of the theory. As Nobel laureate David Gross has mentioned, string theory is like the top of a mountain. As climbers scale the mountain, the top is clearly visible, but it seems to recede the closer you come to it. The goal is tantalizingly close, but seems to be always just out of reach.

Personally, I think this is understandable, since no one knows when, if ever, we will find supersymmetry in the laboratory, but this has to be put into proper perspective. The correctness or incorrectness of a theory should rest on concrete results, not the subjective desires of physicists. We all hope that our pet theories are confirmed within our lifetime. That is a deeply human desire. But sometimes nature has her own timetable.

The atomic theory, for example, took two thousand years before it was finally vindicated, and only recently have scientists been able to take vivid images of individual atoms. Even Newton's and Einstein's great theories took decades for many of their predictions to be fully tested and verified. Black holes were first predicted in 1783 by John Michell, but only in 2019 did astronomers produce the first conclusive pictures of their event horizon.

Personally, I think the pessimism of many scientists might be misguided, because the evidence for the theory might be found not in some gigantic particle accelerator but when someone finds the final mathematical formulation of the theory.

The point here is that perhaps *we do not need an experimental proof of string theory at all.* A theory of everything is also a theory of ordinary things. If we can derive the mass of the quarks and other known subatomic particles from first principles, that might be convincing evidence that this is the final theory.

The problem is not experimental at all. The Standard Model has twenty or so free parameters that are put in by hand (such as the mass of the quarks and the strength of their interactions). We have plenty of experimental data concerning the masses and couplings of subatomic particles. If string theory can precisely calculate these fundamental constants from first principles, without any assumptions, then this would, in my opinion, prove its correctness. It would be a truly historic event if the known parameters of the universe could emerge from a single equation.

But once we have this one-inch-long equation, what do we do with it? How can we escape the landscape problem?

One possibility is that many of these universes are unstable and decay to our familiar universe. We recall that the vacuum, instead of being a boring, featureless thing, is actually teeming with bubble universes popping in and out of existence, like in a bubble bath. Hawking called this the space-time foam. Most of these tiny bubble universes are unstable, jumping out of the vacuum and then jumping back in.

In the same way, once the final formulation of the theory is found, one might be able to show that most of these alternate universes are unstable and decay down to our universe. For example, the natural time scale for these bubble universes is the Planck time, which is 10^{-43} seconds, an incredibly short amount of time. Most universes only live for this brief instant. Yet the age of our universe, by comparison, is 13.8 billion years, which is astronomically longer than the lifespan of most universes in this formulation. In other words, perhaps our universe is special among the infinity of universes in the landscape. Ours has outlasted them all, and that is why we are here today to discuss this question.

But what do we do if the final equation turns out to be so complex that it cannot be solved by hand? Then it seems impossible to show that our universe is special among the universes in the landscape. At that point I think we should put it in a computer. This is the path taken for the quark theory. We recall that the Yang-Mills particle acts like a glue to bind quarks into a proton. But after fifty years, no one has been able to rigorously prove this mathematically. In fact, many physicists have pretty much given up hope of ever accomplishing it. Instead, the Yang-Mills equations are solved on a computer.

This is done by approximating space-time as a series of lattice points. Normally, we think of space-time being a smooth surface, with an infinite number of points. When objects move, they pass through this infinite sequence. But we can approximate this smooth surface with a grid or lattice, like a mesh. As we let the spacing between lattice points get smaller and smaller, it becomes ordinary space-time, and the final theory begins to emerge. Similarly, once we have the final equation for M-theory, we can put it on a lattice and do the computation on a computer.

In this scenario, our universe emerges from the output of a supercomputer.

(However, I am reminded of the *Hitchhiker's Guide to the*

Galaxy, when a gigantic supercomputer is built to find the meaning of life. After eons doing the calculation, the computer finally concluded that the meaning of the universe was "forty-two.")

So it is conceivable that the next generation of particle accelerators, or a particle detector deep inside a mine shaft, or a gravity wave detector in deep space, will find experimental proof of string theory. But if not, then perhaps some enterprising physicist will have the stamina and vision to find the final mathematical formulation of the theory of everything. Only then can we compare it with experiment.

There are probably more twists and turns facing physicists before the journey is finished. But I am sure that we will eventually find the theory of everything.

But the next question is: Where did string theory come from? If the theory of everything has a grand design, then did it have a designer? If so, then does the universe have a purpose and meaning?

7

FIПDIПG MEAПIПG IП THE UПIVERSE

We have seen how the mastery of the four fundamental forces has not only revealed many of the secrets of nature but has also unleashed the great scientific revolutions that have altered the destiny of civilization itself. When Newton wrote down the laws of motion and gravity, he laid the groundwork for the Industrial Revolution. When Faraday and Maxwell revealed the unity of the electric and magnetic force, this set into motion the electric revolution. When Einstein and the quantum physicists revealed the probabilistic and relativistic nature of reality, this set into motion the high-tech revolution of today.

But now we might be converging on a theory of everything that unifies all four fundamental forces. So assume for the moment that we have finally achieved this theory. Assume that it has been rigorously tested and universally accepted by the scientists of the world. What impact will this have on our lives, our thinking, and our conception of the universe?

As far as a direct impact on our immediate lives, it probably will be minimal. Each solution of the theory of everything is an entire universe. Therefore, the energy at which the theory becomes relevant is the Planck energy, which is a quadrillion times greater than the energy produced by the Large Hadron Collider. The energy scale of the theory of everything concerns the creation of the universe and the mysteries of black holes, not the affairs of you and me.

The real impact of the theory on our lives may be philosophical, because the theory may finally answer deep philosophical questions that have haunted great thinkers for generations, such as is time travel possible, what happened before creation, and where did the universe come from?

As the great biologist Thomas H. Huxley said in 1863, "The question of all questions for humanity, the problem which lies behind all others and is more interesting than any of them, is that of the determination of man's place in Nature and his relation to the Cosmos."

But this still leaves open a question: What does the theory of everything have to say about meaning in the universe?

Einstein's secretary, Helen Dukas, once mentioned that Einstein was overwhelmed with the mail he received pleading with him to explain the meaning of life, and asking whether he believed in God. He said he was helpless to answer all these questions about the purpose of the universe.

Today, questions about meaning in the universe and the existence of a creator still fascinate the general public. In 2018, a private letter that Einstein wrote just before he died went up for auction. Surprisingly, the winning bid for the God letter was $2.9 million, even beyond the expectation of the auction house.

In this and other letters, Einstein despaired of answering questions concerning the meaning of life, but he was clear about his thinking concerning God. One problem, he wrote, is that there are really two kinds of Gods, and we often confuse the two. First, there is the personal God, the God that you pray to, the God of the Bible who smites the Philistines and rewards the believers. He did not believe in that God. He did not believe that the God who created the universe interfered in the affairs of mere mortals.

However, he believed in the God of Spinoza—that is, the God of order in a universe that is beautiful, simple, and elegant. The universe could have been ugly, random, chaotic, but instead it has a hidden order that is mysterious yet profound.

As an analogy, Einstein once said he felt like he was a child entering a vast library. All around him, there were stacks of books that contained answers to the mysteries of the universe. His goal in life, in fact, was to be able to read a few chapters of these books.

However, he left open this question: If the universe is like a vast library, is there a librarian? Or is there someone who authored these books? In other words, if all physical laws can be explained by the theory of everything, then where did that equation come from?

And Einstein was driven by another question: Did God have a choice in creating the universe?

PROVING GOD'S EXISTENCE

These questions, however, are not so clear when trying to use logic to prove or disprove the existence of God. Hawking, for example, did not believe in God. He wrote that the Big Bang took place in a brief instant of time, so there was simply not enough time for God to create the universe as we see it.

In Einstein's original theory, the universe expanded almost instantly. But in the multiverse theory, our universe is nothing but a bubble coexisting with other bubble universes, which are being created all the time.

If so, then perhaps time did not simply spring into existence with the Big Bang, but instead there was a time before the beginning of our universe. Each universe was born in a brief instant of time, but the totality of universes in the multiverse could be eternal. So the theory of everything leaves open the question of the existence of God.

Over the centuries, however, theologians have tried the opposite point of view, to use logic to prove the existence of God. Saint Thomas Aquinas, the great Catholic theologian of the thirteenth century, postulated five famous proofs of the existence of God. They are interesting because even today, they still raise deep questions about the theory of everything.

Three are redundant, so there are actually three independent proofs (if we also include the ontological proof of Saint Anselm):

1. Cosmological proof. Things move because they are pushed—that is, something sets them into motion. But what is the First Mover or First Cause that set the universe into motion? This must be God.
2. Teleological proof. Everywhere around us we see objects of great complexity and sophistication. But every design eventually requires a designer. The First Designer was God.
3. Ontological proof. God, by definition, is the most perfect being imaginable. But one can imagine a God that does not exist. But if God did not exist, he would not be perfect. Therefore he must exist.

These proofs of the existence of God lasted for many centuries. It wasn't until the nineteenth century that Immanuel

Kant found a flaw in the ontological proof, because perfection and existence are two separate categories. To be perfect does not necessarily imply that something must exist.

However, the other two proofs have to be reexamined in light of modern science and the theory of everything. The analysis of the teleological proof is straightforward. Everywhere we look around us, we see objects of great complexity. But the sophistication of life-forms surrounding us can be explained by evolution. With enough time, pure chance can drive evolution via the survival of the fittest, so more sophisticated designs arise randomly from less sophisticated designs. A first designer for life is not necessary.

By contrast, the analysis of the cosmological proof is not so clear. Physicists today can run the videotape backward and show that the universe started with a Big Bang that set the universe into motion. However, to go back even before the Big Bang, we have to use the multiverse theory. But even if we assume that the multiverse theory explains where the Big Bang came from, then one has to ask, Where did the multiverse come from? Finally, if one states that the multiverse is a logical consequence of the theory of everything, then we have to ask, Where did the theory of everything come from?

At this point, physics stops, and metaphysics begins. Physics says nothing about where the laws of physics themselves come from. So the cosmological proof of Saint

Thomas Aquinas concerning the First Mover or First Cause is left relevant even today.

The key feature of any theory of everything is likely to be its symmetry. But where does this symmetry come from? This symmetry would be a by-product of deep mathematical truths. But where does mathematics come from? On this question, the theory of everything is again silent.

Questions raised by a Catholic theologian eight hundred years ago are still relevant today, despite our enormous progress in understanding the origin of life and the universe.

MY OWN POINT OF VIEW

The universe is a remarkably beautiful, ordered, and simple place. I find it utterly staggering that all the known laws of the physical universe can be summarized on a single sheet of paper.

Contained on the paper is Einstein's theory of relativity. The Standard Model is more complicated, taking up most of the page with its zoo of subatomic particles. They can describe everything in the known universe, from deep inside the proton to the very boundary of the visible universe.

Given the utter brevity of this sheet of paper, it is hard to avoid the conclusion that this was all planned in advance, that its elegant design shows the hand of a cosmic designer. To me, this is the strongest argument for the existence of God.

But the bedrock of our understanding of the world is science, which is ultimately based on things that are testable, reproducible, falsifiable. That is the bottom line. In disciplines like literary criticism, things get more complicated with time. Analysts forever wonder what James Joyce really meant by this or that passage. But physics moves in the opposite direction, becoming simpler and more powerful with time, until everything is a consequence of a handful of equations. I find this remarkable. But scientists are often reluctant to admit that there are some things beyond the realm of science.

For example, it is impossible to prove a negative.

Let's say we want to prove that unicorns don't exist. Although we have scoured most of the Earth's surface and have never seen a unicorn, there is always the possibility that unicorns exist in some undiscovered island or cave. Thus, it is impossible to prove that unicorns don't exist. This means that a hundred years from now, people will still be debating the existence of God and the meaning of the universe. This is because these concepts are not testable, and hence not decidable. They are outside the province of ordinary science.

Similarly, even if we have never encountered God in all our travels in outer space, there is always the chance that God exists in regions we have never explored.

Hence, I am an agnostic. We have just scratched the surface of the universe, and it is presumptuous to make declarations of the nature of the entire universe far beyond our instruments.

But one still has to confront Saint Thomas Aquinas's proof, that there must be a First Mover. In other words, where did everything come from? Even if the universe started according to the theory of everything, then where did the theory of everything come from?

I believe that the theory of everything exists because it is the only theory that is mathematically consistent. All other theories are inherently flawed and inconsistent. I believe that if you start with an alternate theory, then ultimately you can prove that $2 + 2 = 5$—that is, these alternate theories contradict themselves.

We recall that there is a blizzard of obstacles to a theory of everything. When we add quantum corrections to a theory, we find that the theory usually blows up, with infinite divergences, or the original symmetry is ruined by anomalies. I believe that there is perhaps just one solution to these constraints that fixes the theory, ruling out all other possibilities. The universe cannot exist in fifteen dimensions, since such a universe would suffer from these fatal flaws. (In ten-dimensional string theory, when we calculate quantum corrections, they often contain the term $(D - 10)$, where D is the dimensionality of space-time. Obviously, if we set $D = 10$, then these worrisome anomalies disappear. But if we don't set $D = 10$, then we find an alternate universe full of contradictions, where mathematical logic is violated.

Likewise, when you add in membranes and calculating with M-theory, we find unwanted terms that contain the factor $(D - 11)$. Hence, within string theory, there is only one self-consistent universe where $2 + 2 = 4$, and that is in ten or eleven dimensions.)

This then is a possible answer to the question raised by Einstein in his search for the theory of everything: Did God have a choice in making the universe? Is the universe unique, or are there many ways in which a universe might exist?

If my thinking is correct, then there is no choice. There is only one equation that can describe the universe, because all others are mathematically inconsistent.

So the final equation of the universe is unique. There might be an infinite number of solutions of this master equation, giving us a landscape of solutions, but the equation itself is unique.

This sheds some light on another question: Why is there something rather than nothing?

In the quantum theory, there is no such thing as absolute nothing. We have seen that absolute blackness does not exist, so black holes are really gray and must evaporate. Similarly, when solving the quantum theory, we find that the lowest energy is not zero. For example, you cannot reach absolute zero, because atoms, in their lowest quantum energy state, are still vibrating. (Similarly, according to quantum mechanics, you cannot reach zero energy quantum mechanically, because you still have zero point energy—that is, the lowest quantum vibrations. A state of zero vibration would violate the uncertainty principle, since zero energy is a state of zero uncertainty, which is not allowed.)

So where did the Big Bang come from? Most likely, it was a quantum fluctuation in Nothing. Even Nothing, or a pure vacuum, is frothing with matter and antimatter particles continually jumping out of the vacuum and then collapsing back into the vacuum. This is how something came from nothing.

Hawking, as we saw, called this the space-time foam—that is, a foam of tiny bubble universes continually popping up and disappearing back into the vacuum. We never see this space-time foam, because each bubble is much tinier than any atom. But once in a while, one of these bubbles does not disappear back into the vacuum but continues to expand, until it inflates and creates an entire universe.

So why is there something rather than nothing? Because our universe originally came from quantum fluctuations in Nothing. Unlike countless other bubbles, our universe jumped out of the space-time foam and kept on expanding.

DID THE UNIVERSE HAVE A BEGINNING OR NOT?

Will this theory of everything give us the meaning of life? Years ago, I saw a strange poster from a meditation society. I recognized that it faithfully published all the details of the supergravity equations, in their full mathematical glory. Attached to each term of the equation, however, there was an arrow that said "peace," "tranquility," "unity," "love," etc.

In other words, the meaning of life was embedded in the equations of the theory of everything.

Personally, I think it is unlikely that a purely mathematical term in an equation from physics can be equated to love or happiness.

However, I do believe that the theory of everything might have something to say about meaning in the universe. As a child, I was raised as a Presbyterian, but my parents were Buddhists. These two great religions have, in turn, two diametrically opposed points of view concerning the Creator. In the Christian church, there was an instant of time when God created the world. The Catholic theologian and physicist Georges Lemaître, one of the architects of the Big Bang theory, believed that Einstein's theory was compatible with Genesis.

However, in Buddhism, there is no God. The universe had no beginning or end. There is only timeless Nirvana.

So how can one resolve these two diametrically opposite points of view? The universe either had a beginning. Or it didn't. There is no middle ground.

But actually, the multiverse theory gives a radically new way of viewing this contradiction.

Perhaps our universe did have a beginning, as mentioned in the Bible. But perhaps Big Bangs are happening all the time, according to the inflation theory, creating a bubble bath of universes. Perhaps these universes are expanding in a much larger arena, a Nirvana of hyperspace. So our universe had a beginning and is a three-dimensional bubble floating in a much larger space of eleven-dimensional Nirvana in which other universes continually arise.

Thus, the multiverse idea allows one to combine both the creation mythology of Christianity with the Nirvana of Buddhism into a single theory that is compatible with known physical laws.

MEANING IN A FINITE UNIVERSE

In the end, I believe that we create our own meaning in the universe.

It is too simple and easy to have some guru come down from the mountaintop, bearing the meaning of the universe. The meaning of life is something that we have to struggle to understand and appreciate. Having it given to us defeats the whole purpose of meaning. If the meaning of life were available for free, then it would lose its meaning. Everything that has meaning is the result of struggle and sacrifice, and is worth fighting for.

But it is hard to argue that the universe has a meaning if the universe itself will eventually die. Physics, in some sense, has a death warrant for the universe.

Despite all learned discussions about meaning and purpose in the universe, perhaps it is all for naught, because the universe is doomed to die in a Big Freeze. According to the second law of thermodynamics, everything in a closed system must eventually decay, rust, or fall apart. The natural order of things is to decline and eventually cease to exist. It seems inescapable that all things must die when the universe itself dies. So whatever meaning we may ascribe to the universe will eventually be wiped away when the universe itself dies.

But once again, perhaps the merger of the quantum theory with relativity provides an escape clause. We said that the second law of thermodynamics eventually dooms the universe in a closed system. The key word is *closed*. In an open universe, where energy can enter from the outside, it is possible to reverse the second law.

For example, an air conditioner seems to violate the second law because it takes in chaotic hot air and cools it down. But an air conditioner gets energy from the outside, from a pump, and hence is not a closed system. Likewise, even life on Earth seems to violate the second law, because it takes just nine months to convert hamburgers and french fries into a baby, which truly is a miracle.

So why is life possible on the Earth? Because we have an external source of energy, the sun. The Earth is not a closed system, so sunlight allows us to extract energy from the sun to create the food necessary to feed a baby. So the second law of thermodynamics has an escape clause. Sunlight makes evolution to higher forms possible.

In the same way, it is possible to use wormholes to open a gateway to another universe. Our universe appears to be closed. But one day, perhaps facing the death of the universe, our descendants may be able to use their formidable scientific know-how to channel enough positive energy to open a tunnel through space and time, and then use negative energy (from the quantum Casimir effect) to stabilize the gateway. One day, our descendants will master the Planck energy, the energy at which space and time become unstable, and use their powerful technology to escape our dying universe.

In this way, quantum gravity, instead of being an exercise in the mathematics of eleven-dimensional space-time, becomes a cosmic interdimensional lifeboat allowing intelligent life to evade the second law of thermodynamics and escape to a much warmer universe.

So the theory of everything is more than just a beautiful mathematical theory. Ultimately, it could be our only salvation.

CONCLUSION

The search for the theory of everything has led us into a quest to find the ultimate unifying symmetry of the universe. From the warmth of a summer breeze to the glory of a blazing sunset, the symmetry we see all around us is a fragment of the original symmetry found at the beginning of time. That original symmetry of the superforce was broken at the instant of the Big Bang, and we see remnants of that original symmetry wherever we admire the beauty of nature.

I like to think that perhaps we are like two-dimensional Flatlanders living in some mythical flat plane, unable to visualize the third dimension, which is considered just a superstition. In the beginning of time in Flatland, there was once a beautiful three-dimensional crystal that, for some reason, was unstable and shattered into a million pieces that rained down on Flatland. For centuries, the Flatlanders have tried to reassemble these pieces like a jigsaw puzzle. Over time, they were able to assemble them into two gigantic pieces. One piece was called gravity, the other piece was called the quantum theory. Try as they might, the Flatlanders could never fit these two pieces together. Then one day, an enterprising Flatlander made an outrageous conjecture that set everyone laughing. Why not, he said, using mathematics, lift one of the pieces into an imaginary third dimension so they can fit together, one on top of the other? When this was done, the Flatlanders were amazed and astonished at the dazzling, shimmering jewel that suddenly emerged before them, with its perfect, glorious symmetry.

Or, as Stephen Hawking wrote,

If we do discover a complete theory, it should in time be understandable in broad principle by everyone, not just a few scientists. Then we shall all, philosophers, scientists, and just ordinary people, be able to take part in the discussion of the question of why it is that we and the universe exist. If we find the answer to that, it would be the ultimate triumph of human reason— for then we would know the mind of God.

ACKNOWLEDGMENTS

In writing this book, I am deeply in debt to my agent, Stuart Krichevsky, who has been faithfully at my side for all these decades, giving me sound and wise advice. I always trust his judgment and his intimate understanding of both literary and scientific matters.

I also would like to thank my editor, Edward Kastenmeier, who has guided several of my books with his firm hand and sharp insight. He was the one who suggested that I write this book, and has shepherded the book through all its various stages. This book would have been impossible without his thoughtful and honest advice.

I would also like to thank my colleagues, associates, and friends in the scientific field. In particular, I would like to thank the following Nobel laureates for generously giving me their time and deep insights into physics and the sciences: Murray Gell-Mann, David Gross, Frank Wilczek,

Steve Weinberg, Yoichio Nambu, Leon Lederman, Walter Gilbert, Henry Kendall, T. D. Lee, Gerald Edelman, Joseph Rotblat, Henry Pollack, Peter Doherty, and Eric Chivian. Lastly, I would like to thank the more than four hundred physicists and scientists with whom I have had the pleasure of interacting with, both as collaborators in string research, as well as through my weekly science radio programs, the various TV programs that I have hosted for BBC-TV and the Discovery and Science Channels, and my work as the science correspondent for CBS-TV.

For a more complete list of the scientists whom I have had the pleasure of interviewing, please see my book *The Physics of the Future*. For a more complete list of prominent string theorists whose work I reference in this book, see my Ph.D.-level textbook *Introduction to String Theory and M-Theory*.

NOTES

2   But many others have also tried: In the past, many of the giants
of physics have tried to create their own unified field theory and
failed. In retrospect, we see that a unified field theory must satisfy
three criteria:

    1. It must include all of Einstein's theory of general relativity.
    2. It must include the Standard Model of subatomic particles.
    3. It must yield finite results.

Erwin Schrödinger, one of the founders of the quantum theory,
had a proposal for the unified field theory that was actually studied earlier by Einstein. It failed because it did not reduce to Einstein's theory correctly and could not explain Maxwell's equations.
(It also lacked any description of electrons or atoms.)

Wolfgang Pauli and Werner Heisenberg also proposed a unified field theory that included fermion matter fields, but it was not
renormalizable and did not incorporate the quark model, which
would come decades later.

Einstein himself investigated a series of theories that ultimately
failed. Basically, he tried to generalize the metric tensor for gravity and the Christoffel symbols to include antisymmetric tensors,

in an attempt to include Maxwell's theory in his own theory. This ultimately failed. Simply expanding the number of fields in Einstein's original theory was not enough to explain Maxwell's equations. This approach also made no mention of matter.

Over the years, there have been a number of attempts to simply add matter fields to Einstein's equations, but they have been shown to diverge at the one-loop quantum level. In fact, computers have been used to calculate the scattering of gravitons at the one-loop quantum level, and it has been shown to be conclusively infinite. So far, the only known way to eliminate these infinities at the lowest one-loop level is to incorporate supersymmetry.

A more radical idea was proposed as early as 1919 by Theodor Kaluza, who expressed Einstein's equations in five dimensions. Remarkably, when one curls one dimension into a tiny circle, one finds the Maxwell field coupled to Einstein's gravity field as a result. This approach was studied by Einstein but was eventually abandoned because no one understood how to collapse one dimension. More recently, this approach has been incorporated into string theory, which collapses ten dimensions to four dimensions and in the process generates the Yang-Mills field. So of the many approaches made for a unified field theory, the only path that survives today is the Kaluza higher-dimensional approach, but generalized to include supersymmetry, superstrings, and supermembranes.

More recently, there is a theory called loop quantum gravity. It investigates Einstein's original four-dimensional theory in a new way. However, it is a theory of pure gravity, without any electrons or subatomic particles, and hence cannot qualify as a unified field theory. It makes no mention of the Standard Model, because it lacks matter fields. Also, it is not clear if the scattering of multiloops in this formalism is truly finite. There is speculation that the collision between two loops yields divergent results.

CHAPTER 1: UNIFICATION—THE ANCIENT DREAM

11   "It is with Isaac Newton": Steven Weinberg, *Dreams of a Final Theory* (New York: Pantheon, 1992), 11.

15 So the equations of Newton: Because Newton's *Principia* was
written in a purely geometric fashion, it is clear that Newton was
aware of the power of symmetry. It is also clear that he exploited
the power of symmetry intuitively to calculate the motion of the
planets. However, because he did not use the analytic form of calculus, which would involve symbols like $X^2 + Y^2$, his manuscript
does not represent symmetry analytically in terms of coordinates
X and Y.

23 "We can scarcely avoid": Quotefancy.com, https://quotefancy.com
/quote/1572216/James-Clerk-Maxwell-We-can-scarcely-avoid-the
-inference-that-light-consists-in-the-transverse-undelations-of
-the-same-medium-which-is-the-cause-of-electric-and-magnetic
-phenomena.

24 "So the symmetry": Technically speaking, Maxwell's equations are
not perfectly symmetrical between electric and magnetic fields.
For example, electrons are the sources of electric fields, but Maxwell's equations predict the presence of sources for the magnetic
field as well, called monopoles (i.e., isolated north and south poles
of magnetism), which have never been seen. Therefore, some
physicists have conjectured that these monopoles may eventually
be discovered.

CHAPTER 2: EINSTEIN'S QUEST FOR UNIFICATION

33 "I am nothing but": Abraham Pais, *Subtle Is the Lord* (New York:
Oxford University Press, 1982), 41.

34 "A storm broke loose": Quotation.io, https://quotation.io/page
/quote/storm-broke-loose-mind.

35 "I owe more to Maxwell": Albrecht Fölsing, *Albert Einstein*, trans.
and abridged Ewald Osers (New York: Penguin Books, 1997), 152.

37 "mathematician's patterns": Wikiquotes.com, https://en.wikiquote
.org/wiki/G._H._Hardy.

38 This means that the three: So although special relativity has a four-
dimensional symmetry, as seen by the simple four-dimensional
Pythagorean theorem $X^2 + Y^2 + Z^2 - T^2$ (in certain units), time
enters with an extra minus sign compared to the other spatial
dimensions. This means that time is indeed the fourth dimension,

but of a special type. In particular, it means you cannot easily go back and forth in time (otherwise time travel would be commonplace). One easily goes back and forth in space, but not easily in time, because of this extra minus sign. (Also, notice that we have set the speed of light to be 1, in certain units, to make it clear that time enters into special relativity as the fourth dimension.)

40 "As an older friend": Brandon R. Brown, "Max Planck: Einstein's Supportive Skeptic in 1915," *OUPblog,* Nov. 15, 2015, https://blog .oup.com/2015/11/einstein-planck-general-relativity.

46 "For some days": Fölsing, *Albert Einstein,* 374.

47 "as if I had been wandering": Denis Brian, *Einstein* (New York: Wiley, 1996), 102.

47 "A new scientific truth does not": Johann Ambrosius and Barth Verlag (Leipzig, 1948), p. 22, in Scientific Autobiography and other papers.

49 "Everyone who had any substantial contact": Jeremy Bernstein, "Secrets of the Old One—II," *New Yorker,* March 17, 1973, 60.

CHAPTER 3: RISE OF THE QUANTUM

66 "I think I can safely say": https://en.wikiquote.org/wiki/Talk: Richard_Feynman.

68 "I will never forget the sight": quoted in Albrecht Fölsing, *Albert Einstein,* trans. and abridged Ewald Osers (New York: Penguin Books, 1997), 516.

68 "It was the greatest debate": quoted in Denis Brian, *Einstein* (New York: Wiley, 1996), 306.

70 With the success of quantum theory: Even today, there is no universally accepted solution to the cat problem. Most physicists simply use quantum mechanics as a cookbook that always yields the proper answer and ignore the subtle, deep philosophical implications. Most graduate courses on quantum mechanics (including the one that I teach) simply mention the cat problem but offer no definitive solution. Several solutions have been proposed, which are usually variations of two popular approaches. One is to acknowledge that the consciousness of the observer has to be part of the measuring process. There are variations to this approach,

depending on how you define "consciousness." Another approach, which is gaining popularity among physicists, is the multiverse theory, where the universe splits in half, with one universe containing the live cat, and another containing a dead cat. It is, however, nearly impossible to go back and forth between these two universes, because they have "decohered" from each other—that is, they no longer vibrate in unison, so they can no longer communicate with each other. In the same way that two radio stations cannot interact with each other, we have decohered from all the other parallel universes. So bizarre quantum universes might coexist with ours, but communicating with them is almost impossible. We might have to wait longer than the lifetime of the universe to pass into these parallel universes.

CHAPTER 4: THEORY OF ALMOST EVERYTHING

78 "You are on a lion hunt": Denis Brian, *Einstein* (New York: Wiley, 1996), 359.

78 "I believe I am right": quoted in Walter Moore, *A Life of Erwin Schrödinger* (Cambridge: Cambridge University Press, 1994), 308.

79 "We in the back": Nigel Calder, *The Key to the Universe* (New York: Viking, 1977), 15.

79 "It was an uncanny encounter": quoted in William H. Cropper, *Great Physicists* (Oxford: Oxford University Press, 2001), 252.

81 "The numerical agreement": Steven Weinberg, *Dreams of a Final Theory* (New York: Pantheon, 1992; New York: Vintage, 1994), 115.

82 "This is just not sensible mathematics": John Gribbin, *In Search of Schrödinger's Cat* (New York: Bantam Books, 1984), 259.

90 "if I had known": quoted in Dan Hooper, *Dark Cosmos* (New York: HarperCollins, 2006), 59.

93 "I have committed": Frank Wilczek and Betsy Devine, *Longing for Harmonies* (New York: Norton, 1988), 64.

95 Physicist Sheldon Glashow would exclaim: Robert P. Crease and Charles C. Mann, *The Second Creation* (New York: Macmillan, 1986), 326.

99 They realized that by cobbling together three theories: The mathematical symmetry that mixes three quarks is called SU(3), the

special unitary Lie group of degree 3. So by rearranging the three quarks according to the symmetry SU(3), the final equation for the strong nuclear force must remain the same. The symmetry that mixes the electron and neutrino in the weak nuclear force is called SU(2), the Lie group in degree 2. (In general, if we start with n fermions, then it is straightforward to write down a theory with SU(n) symmetry.) The symmetry coming from Maxwell's theory is called U(1). Therefore, by gluing these three theories together, we find that the Standard Model has symmetry SU(3) × SU(2) × U(1).

Although the Standard Model fits all the experimental data on subatomic physics, the theory seems contrived, because it is based on mechanically patching three forces together.

101 Second, the Standard Model: To compare the simplicity of Einstein's equations to the complexity of the Standard, we note that Einstein's theory can be summarized in just a short equation:

$$G_{\mu\nu} \equiv R_{\mu\nu} - \frac{1}{2}Rg_{\mu\nu} = \frac{8\pi G}{c^4}T_{\mu\nu}$$

while the Standard Model's equations (in highly abbreviated form) require most of the page to write, detailing the various quarks, electrons, neutrinos, gluons, Yang-Mills particles, and Higgs particles.

$$\mathcal{L} = -\frac{1}{2}\mathrm{Tr}\,G_{\mu\nu}G^{\mu\nu} - \frac{1}{2}\mathrm{Tr}\,W_{\mu\nu}W^{\mu\nu} - \frac{1}{4}F_{\mu\nu}F^{\mu\nu}$$

$$+ (D_\mu\phi)^\dagger D^\mu\phi + \mu^2\phi^\dagger\phi - \frac{1}{2}\lambda\left(\phi^\dagger\phi\right)^2$$

$$+ \sum_{f=1}^{3}(\bar{\ell}_L^f i\displaystyle{\not}D\ell_L^f + \bar{\ell}_R^f i\displaystyle{\not}D\ell_R^f + \bar{q}_L^f i\displaystyle{\not}D q_L^f + \bar{d}_R^f i\displaystyle{\not}D d_R^f + \bar{u}_R^f i\displaystyle{\not}D u_R^f)$$

$$- \sum_{f=1}^{3} y_\ell^f(\bar{\ell}_L^f \phi \ell_R^f + \bar{\ell}_R^f \phi^\dagger \ell_L^f)$$

$$- \sum_{f,g=1}^{3} \left(y_d^{fg}\bar{q}_L^f \phi d_R^g + (y_d^{fg})^*\bar{d}_R^g \phi^\dagger q_L^f + y_u^{fg}\bar{q}_L^f \tilde{\phi} u_R^g + (y_u^{fg})^*\bar{u}_R^g \tilde{\phi}^\dagger q_L^f\right)$$

Remarkably, we know that all physical laws of the universe can, in principle, be derived from this one page of equations. The problem is that the two theories—Einstein's relativity theory and the Standard Model—are based on different mathematics, different assumptions, and different fields. The ultimate goal is to merge these two sets of equations into a single, finite unified fashion. The key observation is that any theory claiming to be the theory of everything must contain both sets of equations, yet remain finite. So far, of all the various theories that have been proposed, the only theory that can do this is string theory.

CHAPTER 6: RISE OF STRING THEORY: PROMISE AND PROBLEMS

143 My colleague Keiji Kikkawa: Dr. Kikkawa and I are cofounders of a branch of string theory called "string field theory," which allows us to express the sum total of string theory in the language of fields, resulting in a simple equation a bit over one inch long:

$$L = \Phi^\dagger \left(i\partial_\tau - H \right) \Phi + \Phi^\dagger * \Phi * \Phi$$

Although this allows us to express all of string theory in compact form, it is not the final formulation of the theory. As we shall see, there are five different types of string theory, each requiring a string field theory. But if we go to the eleventh dimension, all five theories apparently converge into one equation, described by something called M-theory, which includes a variety of membranes as well as strings. At present, because membranes are so hard to work with mathematically, especially in eleven dimensions, no one has been able to express M-theory in a single field theory equation. This, in fact, is one of the major goals of string theory: to find the final formulation of the theory from which we can extract physical results. In other words, string theory is probably not yet in its final form.

151 "Although the symmetries are hidden": quoted in Nigel Calder, *The Key to the Universe* (New York: Viking, 1977), 185.

155 Soon after M-theory was proposed: More precisely, the duality found by Maldacena was between $N = 4$ supersymmetric Yang-Mills theory in four dimensions and type IIB string theory in ten dimensions. This is a highly nontrivial duality, because it shows the equivalence between a gauge theory with Yang-Mills particles

in four dimensions and string theory in ten dimensions, which are usually thought to be distinct. This duality showed the deep relationship between gauge theories, which are found in the strong interactions in four dimensions, and ten-dimensional string theory, which is remarkable.

159 "What you said was so confused": quoted in William H. Cropper, *Great Physicists* (New York: Oxford University Press, 2001), 257.

160 "Years of intense effort": http://www.preposterousuniverse.com /blog/2011/10/18/column-welcome-to-the-multiverse/comment -page-2.

161 "I find myself a dinosaur": Sheldon Glashow, with Ben Bova, *Interactions* (New York: Warner Books, 1988), 330.

162 "The research worker": quoted in Howard A. Baer and Alexander Belyaev, *Proceedings of the Dirac Centennial Symposium* (Singapore: World Scientific Publishing, 2003), 71.

162 "Beautiful theories have been": Sabine Hossenfelder, "You Say Theoretical Physicists Are Doing Their Job All Wrong. Don't You Doubt Yourself?," *Back Reaction* (blog), Oct. 4, 2018, http://back reaction.blogspot.com/2018/10/you-say-theoretical-physicists -are.html.

CHAPTER 7: FINDING MEANING IN THE UNIVERSE

198 "If we do discover a complete theory": Stephen Hawking, *A Brief History of Time* (New York: Bantam Books, 1988), 175.

SELECTED READING

Bartusiak, Marcia. *Einstein's Unfinished Symphony.* Yale University Press, 2017.

Becker, Katrin, Melanie Becker, and John Schwarz. *String Theory and M-Theory.* Cambridge University Press, 2007.

Crease, Robert P., and Charles Mann. *The Second Creation: Makers of the Revolution in Twentieth-Century Physics.* New York: Macmillan, 1986.

Einstein, Albert. *The Special and General Theory.* Mineola, New York: Dover Books, 2001.

Feynman, Richard. *Surely You're Joking, Mr. Feynman: Adventures of a Curious Character.* New York: W. W. Norton, 2018.

———. *The Feynman Lectures on Physics* (with Robert Leighton and Matthew Sands). New York: Basic Books, 2010.

Green, Michael, John Schwarz, and Edward Witten. *Superstring Theory,* vols. 1 and 2. Cambridge: Cambridge University Press, 1987.

Greene, Brian. *The Elegant Universe: Superstrings, Hidden Dimensions, and the Quest for the Ultimate Theory.* New York: W. W. Norton, 2010.

Hawking, Stephen. *A Brief History of Time.* New York: Bantam, 1998.

———. *The Grand Design* (with Leonard Mlodinow). New York: Bantam, 2010.

Hossenfelder, Sabine. *Lost in Math: How Beauty Leads Physics Astray.* New York: Basic Books, 2010.

Isaacson, Walter. *Einstein: His Life and Universe.* New York: Simon and Schuster, 2008.

Kaku, Michio. *Parallel Worlds: A Journey Through Creation, Higher Dimensions, and the Future of the Cosmos.* New York: Random House. 2006.

———. *Hyperspace: A Scientific Odyssey Through Parallel Universes, Time Warps, and the Tenth Dimension.* New York: Oxford University Press, 1995.

———. *Introduction to String Theory and M-Theory.* New York: Springer-Verlag, 1999.

Kumar, Manhit. *Quantum: Einstein, Bohr, and the Great Debate About the Nature of Reality.* New York: W. W. Norton, 2010.

Lederman, Leon. *The God Particle: If the Universe Is the Answer, What Is the Question?* New York: Mariner Books, 2012.

Levin, Janna. *Black Holes Blues and Other Songs from Outer Space.* New York: Anchor Books, 2017.

Maxwell, Jordan. *The History of Physics: The Story of Newton, Feynman, Schrodinger, Heisenberg, and Einstein.* Independently published, 2020.

Misner, Charles W., Kip Thorne, and John A. Wheeler. *Gravitation.* Princeton: Princeton University Press. 2017.

Mlodinow, Leonard. *Stephen Hawking: A Memoir of Friendship and Physics.* New York: Pantheon Books, 2020.

Polchinski, Joseph. *String Theory,* vols. 1 and 2. Cambridge: Cambridge University Press, 1999.

Smolin, Lee. *The Trouble with Physics: The Rise of String Theory, the Fall of a Science, and What Comes Next.* New York: Houghton Mifflin, 2006.

Thorne, Kip. *Black Holes and Time Warps: Einstein's Outrageous Legacy.* New York: W. W. Norton, 1994.

Tyson, Neil de Grasse. *Death by Black Hole and Other Cosmic Quandaries.* New York: W. W. Norton, 2007.

Weinberg, Steven. *Dreams of a Final Theory: The Scientific Search for the Ultimate Laws of Nature.* New York: Vintage Books, 1992.

Wilczek, Frank. *Fundamentals: Ten Keys to Reality.* New York: Penguin Books, 2021.

Woit, Peter. *Not Even Wrong: The Failure of String Theory and the Search for Unity in Physical Law.* New York: Basic Books, 2006.

INDEX

quantum mechanics *(continued)*
  X-ray crystallography and,
    85–86
  zero energy and, 191–92
  see also Standard Model
quantum radiation, emitted by
    black holes, 114
quarks, 3, 90–91, 94, 101–2, 203n
  three-quark symmetry and, 91,
    94, 96, 99, 208n
  Yang-Mills theory and, 94–96,
    180, 204n

radio, invention of, 3, 25–26
radioactive decay, 54, 92
radium, 53–55, 88–89
relativity theory, 32–49, 182, 188,
    208–9n
  confirmation of, 45–49
  four-dimensional symmetry
    of, 36–39, 50, 63, 77–78, 91,
    204n, 205–6n
  unification of space and time
    and matter and energy in, 36
  see also general relativity;
    special relativity
Renaissance, 9–10, 71, 107
renormalization, 82, 91, 103–4, 150
Rosen, Nathan, 119
Rutherford, Ernest, 21, 54–55,
    88–89

Sagan, Carl, 159
Sakita, Bunji, 148
Salam, Abdus, 94
satellites, 84, 131, 170, 174
  GPS and, 48–49

microwave background
    radiation detected by,
    132–33, 168–69
Schrödinger, Erwin, 3–4, 66, 67,
    69–70, 74
  life force and, 84–85
  unified field theory of, 78, 203n
Schrödinger equation, 60–64
  Dirac's theory of electron and,
    63–65
  limitations of, 63–64
  periodic table and, 61–62
Schrödinger's cat, 69–70,
    206–7n
Schwarz, John, 144–45
Schwarzschild, Karl, 109–10, 117
Schwinger, Julian, 80–81, 139
Shakespeare, William, 43, 67
singularities, 117, 121
Snyder, Hartland, 111
solar eclipse, sun's gravity
    during, 46
Solvay Conference (1930), 67–68,
    75, 159–60
space-time, 38–39, 41–42, 50, 63,
    91, 180
space-time foam, 179, 192
sparticles, 149
special relativity, 35, 39–40, 45, 50,
    63, 77–78, 80, 82, 205–6n
  Einstein's insights leading to,
    32–35, 39
  GPS and, 48–49
spectrographs, 72–73
speed of light, 23, 63, 119, 169
  escape velocity and, 108–9
  relativity theory and, 34–35